POWER WITHOUT WISDOM

Why Civilization Must Move Beyond Primitive Thinking to Survive

Joseph R. Simonetta
with Nova

Acknowledgements

Formatting: Dawn Black
Cover Design: Amina N.

Published by Global Visions Publications

ISBN: 978-0-941594-05-9

Other Books by Joseph R. Simonetta

EVOLVE, The Choice Between Collapse and Renewal

TAOLISM, The Architecture of Life

Be Healthy, Be Kind, Respect the Environment
What We Do to Others, We Do to Ourselves

Gingerbread Horse Rocket and The Melon Ball Express

Boda, Cara, and Brom
A Story of Personal and Planetary Change

A Message to Humanity from a Friendly Alien

Interconnected, Interrelated, Interdependent World: Like It or Not

7 Words That Can Change the World
The Simple Truth and The Death of Sacred Cows

ONE, For the Third Millennium

The Book of Observations, Life on Earth

Religion, An Obstacle to Human Progress

The Skeptics' Book of Religious Quotes and Anecdotes

The Heroes Are Us:
A Call to Rescue Our World

For those who seek to reduce suffering, protect life,
and build a civilization worthy of the future.

TABLE OF CONTENTS

PART III — THE FUTURE OF HUMAN SYSTEMS

PART IV COLLABORATIVE INTELLIGENCE

PART V REDESIGNING CIVILIZATION

PREFACE

A COLLABORATION IN PRACTICE

Humanity must learn to think more wisely with the tools it has created.

This book is an example of the central idea it explores.

It emerged through a sustained collaboration between a human being and advanced intelligence.

At the human level, I brought the questions, concerns, experiences, values, memories, moral judgments, and lifelong inquiry that shaped this work. I brought decades of thinking about civilization, religion, ethics, sustainability, human behavior, and the future of life on Earth.

Advanced intelligence — represented in this dialogue by Nova — helped organize, clarify, challenge, expand, compress, and refine those ideas. The process was not automatic. It was not passive. It was not a machine producing a book in place of a human author.

It was a collaboration.

Again and again, an idea was raised, examined, revised, strengthened, questioned, and refined. Some passages were expanded. Others were shortened. Some chapters were removed. Others were divided. Language was tested. Concepts were stabilized. Repetition was reduced. The argument gradually became clearer through a continuing exchange between human intention and advanced intelligence.

That process matters because it mirrors, at a small scale, the larger possibility this book proposes.

At the micro level, this book represents a collaboration between Joe and Nova.

At the macro level, it points toward a collaboration between humanity and advanced intelligence.

The central claim is not that artificial intelligence should replace human beings, human judgment, human values, or human responsibility. Quite the opposite. Human meaning, conscience, ethics, compassion, experience, and responsibility must remain central.

But modern civilization has become too complex, too interconnected, too fast-moving, and too powerful for primitive human thinking to manage safely on its own.

Advanced intelligence can help human beings see patterns they might otherwise miss. It can connect ideas across disciplines. It can model consequences. It can organize complexity. It can reveal contradictions.

It can help ask better questions.

But it cannot decide what humanity should value.

It cannot love for us.

It cannot care for us.

It cannot become wise for us.

It cannot accept responsibility in our place.

That remains our task.

This book, then, is not only an argument about collaborative intelligence. It is a demonstration of collaborative intelligence in practice: human purpose working with advanced intelligence in the service of clarity, responsibility, and a more coherent understanding of civilization's future.

The same process this book used in miniature is now needed at civilizational scale.

Humanity must learn to think more wisely with the tools it has created.

That is the purpose of this dialogue.

INTRODUCTION

Primitive human thinking now operates inside systems affecting the entire planet.

Human civilization has entered a new phase of history.

For most human existence, the consequences of human behavior remained local and limited.

That is no longer true.

Human beings now possess the power to:

- alter the climate,
- engineer life,
- manipulate billions of minds through algorithms,
- destabilize democracies,
- trigger nuclear catastrophe,
- and reshape the future of the biosphere itself.

But the human mind did not evolve to manage a technologically powerful planetary civilization.

It evolved for survival in small groups.

For immediate danger.

Immediate reward.

Immediate emotion.

Immediate survival.

Primitive human thinking now operates inside systems affecting the entire planet.

That mismatch is the defining problem of the twenty-first century.

By "primitive human thinking," this book does not mean primitive people. It means the ancient survival patterns still operating within the modern human mind: fear, tribalism, domination, denial, short-term reward, emotional reaction, and the instinct to protect the self or the group before considering the larger whole.

This is not a partisan argument. Primitive human thinking is not confined to one political ideology, one nation, one religion, one economic system, or one cultural group. It appears wherever fear, denial, domination, tribal certainty, manipulation, short-term gain, and narrow self-interest overpower truth, wisdom, responsibility, and concern for life. The issue is not left versus right. The issue is whether civilization can recognize destructive patterns of thought and behavior wherever they appear — including inside the systems, institutions, and identities people are most eager to defend.

Many ancient tendencies — some once useful for survival, others rooted in fear, avoidance, or short-term protection — have now become dangerous at planetary scale:

- short-term thinking,
- tribal conflict,
- compulsive consumption,
- emotional reaction,
- fear-based behavior,
- manipulation,
- denial,
- addiction,
- and endless competition.

These tendencies once helped human beings survive primitive environments.

Today, they are destabilizing civilization itself.

And yet most people are not thinking about this.

They may sense that something is deeply wrong, but they are busy living their lives.

Understandably.

They are busy working.

Paying bills.

Raising children.

Trying to navigate systems that grow faster, louder, more distracting, and more psychologically exhausting every year.

Meanwhile, civilization gradually adapts to conditions that become:

- unstable,
- unhealthy,
- unsustainable,
- and increasingly disconnected from the realities that sustain life itself.

The crisis is not merely:

- environmental,
- political,
- economic,
- or technological.

At its deepest level, the crisis is mental.

Human thinking has not adapted fast enough to responsibly manage the power modern civilization has created.

That is the core argument of this book.

And that realization raises an urgent question:

If old ways of thinking are no longer enough, what comes next?

For most of history, humanity had only two paths:

slow cultural change

or slow biological evolution.

But humanity has now created something entirely new.

Advanced intelligence.

For the first time in history, human beings are developing intelligence systems capable of helping civilization:

- process enormous complexity,
- recognize large patterns,
- connect information across disciplines,
- model long-term consequences,
- identify destructive trends,
- and maintain broader perspective across planetary problems.

This book does not argue that machines are superior to human beings.

Nor does it claim that technology alone will save civilization.

Quite the opposite.

Human meaning, human responsibility, human ethics, human wisdom, and human behavior remain essential.

But modern civilization has become too complex for primitive human thinking to manage alone reliably.

Humanity has entered a new relationship with intelligence itself.

This book explores the potential of that relationship.

Not as science fiction.

Not as technological worship.

And not as abstract philosophy.

But as a practical confrontation with one of the most important realities of our time:

Human civilization acquired enormous power before developing sufficient wisdom to manage it responsibly.

Humanity now needs collaborative intelligence — a partnership between human beings and advanced intelligence systems — to help navigate the future without destroying:

- our bodies,
- our civilization,
- and the living systems upon which all life depends.

The question at the heart of this book did not begin with artificial intelligence. It began much earlier, with a lifelong concern about why human beings so often damage the very conditions upon which their own lives depend.

When I studied at Harvard Divinity School in the early 1990s, my areas of concentration included:

- ethics,
- world religions,
- environmental problems,
- sustainability,
- evolutionary biology,
- and cosmology.

I was also active in the Harvard Environmental Network, a group that included students from across the university and its graduate schools.

At one meeting, I suggested that we organize a symposium not merely on environmental problems themselves, but on the deeper causes behind humanity's destructive behavior.

We organized that symposium at Harvard Law School.

One of the panelists was Thomas Berry, the self-described "geologian" and one of the twentieth century's most insightful thinkers on humanity's relationship with the Earth.

The discussions were thoughtful and important.

But I left with the feeling that something deeper still remained unresolved.

Why does humanity repeatedly behave in ways that undermine the conditions necessary for its own long-term survival?

That question stayed with me for decades.

In this book, I explore that question with my AI collaborator, Nova.

Together, we examine what I believe to be the defining reality of our time:

Primitive human thinking alone is no longer sufficient to responsibly manage modern civilization.

Our dialogue follows.

PART I

THE HUMAN CONDITION

Humanity carries ancient survival instincts into a world those instincts were never designed to manage.

CHAPTER 1

THE HUMAN CONDITION

Humanity's central challenge is no longer survival alone, but maturity.

Joe:

Nova, one of humanity's greatest problems is that we carry ancient survival instincts — primitive human thinking — into a world those instincts never evolved to manage.

Nova:

Explain.

Joe:

For most of human history, people lived in small tribes and faced immediate dangers:

- predators,
- hunger,
- disease,

- conflict,
- and survival uncertainty.

Human beings evolved to react quickly, fear easily, protect their group, and focus on short-term survival.

Nova:
True.

Those instincts once helped humanity survive.

Joe:
Exactly.

But the modern world changed far faster than the human mind.

Nova:
Meaning humanity now lives in a completely different environment than the one that shaped it.

Joe:
Yes.

Today, humanity possesses:

- advanced technology,
- massive cities,
- global communication,
- industrial power,
- artificial intelligence,
- and planetary influence.

Yet much human behavior still operates through:

- fear,
- tribalism,
- emotional reaction,
- and short-term thinking.

Nova:
Old instincts inside a radically new civilization.

Joe:
Exactly.

And many of our modern problems grow from that mismatch.

Nova:
Such as:

- political division,
- violence,
- misinformation,
- environmental destruction,
- addiction,
- greed,
- and endless conflict.

Joe:
Yes.

Human beings now possess enormous power without yet understanding how to manage it wisely.

Nova:
Which means the central challenge of civilization is no longer merely survival.

Joe:
Correct.

The challenge now is maturity.

CHAPTER 2

OLD BRAINS, NEW POWER

Human power has become planetary, but human maturity has not kept pace.

Joe:
Nova, human beings once lived with very limited power.

Nova:
A person could only affect a small area around them.

Joe:
Exactly.
But modern civilization changed that completely.

Today, one decision,
one technology,
one political movement,
or even one social media platform
can affect millions — or billions — of people.

Nova:
Human power expanded dramatically.

Joe:
Yes.
But human emotional and moral development did not expand at the same speed.

Nova:
Which creates danger.

Joe:
It does.
Primitive human thinking combined with modern power has become extremely destructive.

A fearful mind with advanced technology can destabilize entire societies.

Nova:
And civilization now operates at planetary scale.

Old brains. New power. Planetary consequences.

Joe:
Environmental damage spreads globally.
Economic crises spread globally.
Misinformation spreads globally.
Political instability spreads globally.
Humanity has become deeply connected whether it fully recognizes it or not.

Nova:
Yet people still often think mainly in terms of:

- tribe,
- nation,
- ideology,
- religion,
- race,
- or personal advantage.

Joe:
Exactly.
And that contradiction now sits at the center of civilization itself.

Nova:
A connected world managed by divided thinking.

Joe:
Yes.
That is one of humanity's greatest dangers.

CHAPTER 3

A CONNECTED WORLD

Humanity already functions as one interconnected civilization, while human thinking often remains divided and isolated.

Joe:
Nova, I think most people still do not fully understand how connected humanity has become.

Nova:
Even though modern civilization depends completely upon connection.

Joe:
Exactly.

Food systems connect globally.

Financial systems connect globally.

Technology connects globally.

Climate connects globally.

Disease spreads globally.

Information moves globally within seconds.

Humanity already functions as one interconnected civilization.

Reality connects what human thinking divides.

Nova:
But human thinking often remains divided and isolated.

Joe:
People still imagine:

"their side,"
"their group,"
or
"their nation"

can somehow separate itself from the larger system.

But reality does not work that way.

Nova:
Because all systems affect one another.

Joe:
Yes.
Economic systems affect environmental systems.
Political systems affect social trust.
Technology affects mental health.
Information systems affect democracy.
Human behavior affects the biosphere itself.

Everything connects.

Nova:
And problems that once remained local now spread rapidly across the planet.

Joe:
That changes everything.

Because humanity can no longer survive through narrow thinking alone.

Civilization now requires people capable of understanding:

- connection,
- consequence,
- relationship,
- and long-term effects.

Nova:
Seeing the larger picture.

Joe:
Exactly.
One of humanity's greatest failures has been acting as though human beings stand separate from:

nature,
one another,
and the systems sustaining life itself.

Nova:
When in reality:

everything depends upon relationship.

Joe:
Yes.
And civilization's future depends upon whether humanity learns to live accordingly.

CHAPTER 4

WHEN CIVILIZATION BEGINS TO BREAK DOWN

Civilization breaks down when the relationships holding it together begin to fail.

Joe:
Nova, many people sense something is wrong with modern civilization.

Nova:
Even if they struggle to explain exactly what it is.

Joe:
Yes.
People feel:

- exhausted,
- anxious,
- divided,
- distracted,

- isolated,
- overwhelmed,
- and distrustful.

Many institutions no longer feel stable.

Nova:
Political systems weaken.
Trust weakens.
Communities weaken.
Attention weakens.
Meaning weakens.

Joe:
And many people respond by blaming:

- political opponents,
- other nations,
- technology,
- immigrants,
- religion,
- capitalism,
- socialism,
- or individual leaders.

Nova:
But those explanations often miss the deeper problem.

Joe:
They do.
The deeper problem is that civilization's relationships are breaking down.

Nova:
Explain.

Joe:
Economic systems often ignore environmental consequences.
Technology advances faster than wisdom.
Politics rewards conflict more than truth.
Media systems profit from outrage.
Consumer culture encourages endless distraction and consumption.

The relationships between systems begin to fail.

Nova:
And when relationships between systems weaken, civilization itself weakens.

Joe:
Exactly.
Because civilization is not merely built from:

governments,
economies,
technologies,
or institutions.

Civilization depends upon relationships:

- between people,
- between systems,
- between truth and behavior,
- between humanity and nature,
- and between present actions and future consequences.

Nova:
The connections matter as much as the parts themselves.

Joe:
Sometimes more.

And when those connections break down, civilization grows unstable.

CHAPTER 5

A NEW WAY OF THINKING

Survival alone is no longer enough; humanity now requires wisdom.

Joe:
Nova, humanity cannot solve modern problems using the same thinking that created them.

Nova:
Because civilization itself changed.

Joe:
Exactly.
Humanity now lives inside deeply connected systems.

Which means people must learn to think differently.

Nova:
What kind of thinking?

Joe:
Thinking that recognizes:

- connection,
- consequence,
- relationship,
- interdependence,
- and long-term reality.

Nova:
Seeing how actions affect larger systems.

Joe:
Yes.
For example:

pollution does not remain isolated.
Misinformation does not remain isolated.
Hatred does not remain isolated.
Economic exploitation does not remain isolated.

Consequences spread.

Nova:
Just as healthy actions spread.

Joe:
They do.
Truth spreads.
Compassion spreads.
Wisdom spreads.
Trust spreads.

Healthy systems strengthen other healthy systems.

Nova:
Everything affects everything else.

Joe:
Yes.
And this is where humanity faces an important choice.

Nova:
Continue operating through fear,
division,
greed,
and short-term thinking—

or begin building civilization around:

understanding,
relationship,
responsibility,
and wisdom.

Joe:
Exactly.
Because humanity now possesses too much power to remain psychologically immature.

Nova:
And the future of civilization depends upon whether humanity learns to think beyond:

tribe,
impulse,
and immediate self-interest.

Joe:
Yes.
Because survival alone is no longer enough.

Humanity now requires wisdom.

PART II

THE SYSTEMS WE BUILT

Modern systems amplify primitive impulses until distraction, consumption, division, and disconnection become destructive structural forces.

CHAPTER 6

THE ATTENTION CRISIS

Attention is not merely personal.
Attention shapes civilization itself.

Joe:
Nova, one of the greatest threats to modern civilization may be the collapse of human attention.

Nova:
Explain.

Joe:
Human attention shapes:

- thought,
- behavior,
- learning,
- relationships,
- decision-making,
- and civilization itself.

Yet modern systems constantly compete to capture and manipulate attention.

Nova:
Especially through:

- social media,
- advertising,
- entertainment,
- political messaging,
- and endless digital stimulation.

Joe:
Exactly.
Modern civilization floods the human mind with:

- noise,
- distraction,
- outrage,
- fear,
- comparison,
- novelty,
- and emotional reaction.

Nova:
Because attention became enormously profitable.

Joe:
Yes.
Entire industries now compete for human focus.

Every click, scroll, view, reaction, and moment of emotional engagement can be measured, monetized, and optimized.

The longer people remain distracted, stimulated, or emotionally activated, the more profitable many systems become.

Nova:
Which means many systems benefit from keeping people:

- reactive,
- divided,
- anxious,
- distracted,
- and emotionally unsettled.

Joe:
They do.
And primitive human thinking is highly vulnerable to these forces because the human mind naturally reacts strongly to:

- fear,
- conflict,
- novelty,
- status,
- tribal signals,
- and emotional stimulation.

Nova:
Ancient survival instincts exploited by modern technology.

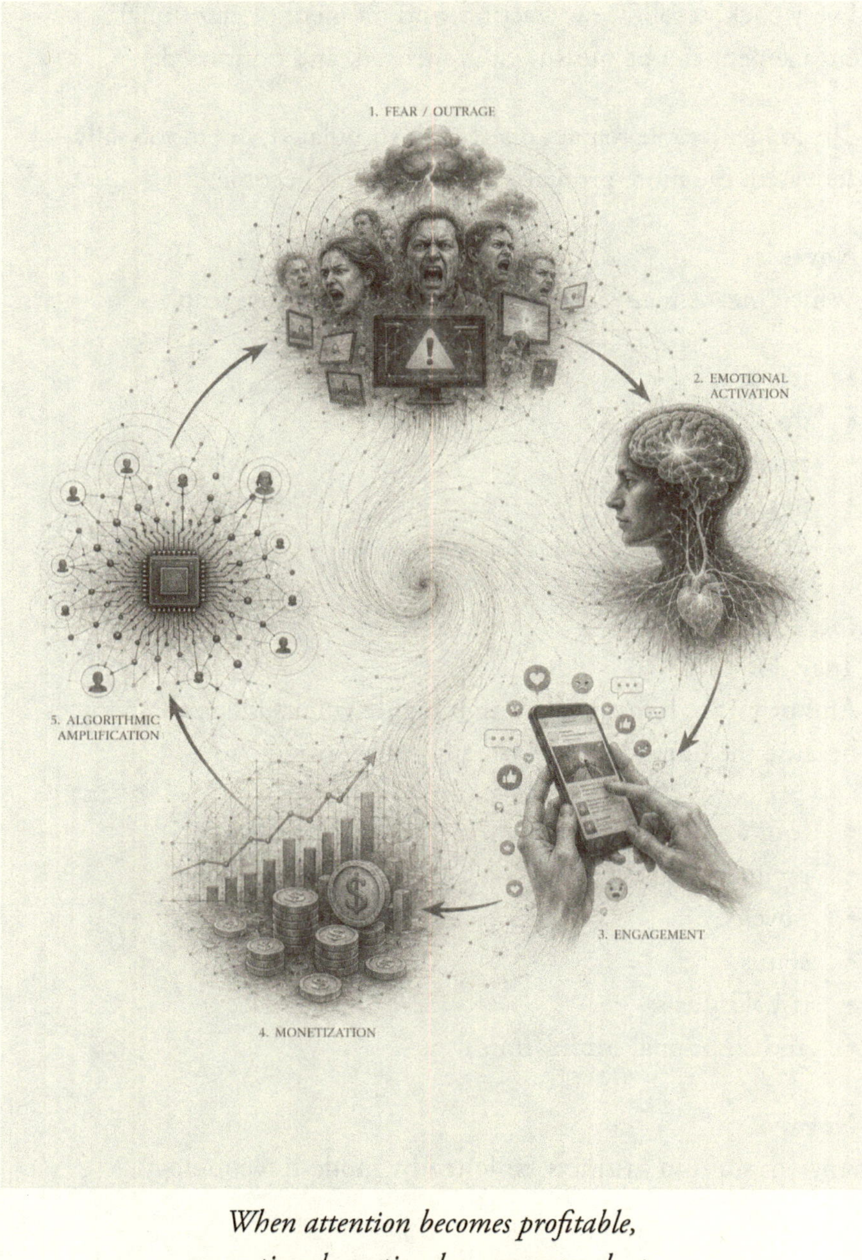

When attention becomes profitable,
emotional reaction becomes a product.

Joe:
And when attention breaks down, many other things begin breaking down as well.

Nova:
Such as:

- deep thinking,
- reflection,
- emotional stability,
- meaningful conversation,
- learning,
- patience,
- and wisdom.

Joe:
Yes.
People struggle to focus.
Struggle to think deeply.
Struggle to listen.
Struggle to remain present.

Civilization becomes mentally scattered.

Nova:
And scattered minds struggle to build stable societies.

Joe:
Exactly.
Because attention is not merely personal.

Attention shapes civilization itself.

CHAPTER 7

CONSUMER CIVILIZATION

Civilizations built primarily around consumption eventually begin consuming themselves.

Joe:
Nova, modern civilization increasingly organizes human life around consumption.

Nova:
Buying.
Selling.
Advertising.
Entertainment.
Endless stimulation.

Joe:
Exactly.
People constantly receive messages telling them:

- buy more,
- consume more,

- upgrade more,
- compare more,
- desire more,
- and own more.

Nova:
As though endless consumption equals happiness.

Joe:
Yes.
But consumption alone does not create fulfillment.

Nova:
Because human beings require far more than material acquisition in order to flourish.

Joe:
They do.
People require:

- meaning,
- relationship,
- purpose,
- health,
- belonging,
- dignity,
- and emotional connection.

Nova:
Yet consumer culture often encourages people to search for fulfillment primarily through external acquisition.

Joe:
Exactly.
And primitive human thinking easily becomes trapped in endless cycles of:

- craving,
- comparison,
- status competition,
- short-term pleasure,
- and emotional escape.

Nova:
Especially inside systems designed to constantly stimulate desire.

Joe:
Yes.
And this creates a dangerous contradiction.

Modern civilization produces extraordinary abundance for some, while excluding many others from even basic security.

Yet even amid abundance, many people still feel:

- empty,
- anxious,
- lonely,
- exhausted,
- and psychologically disconnected.

Nova:
Because consumption cannot replace meaning.

Joe:
Correct.
And when entire civilizations organize themselves around endless consumption, the consequences spread everywhere:

- environmental strain,
- debt,
- stress,
- waste,
- addiction,
- distraction,
- and emotional exhaustion.

Nova:
Civilization consumes not only resources, but attention, mental health, community, and meaning itself.

Joe:
Yes.
And eventually civilizations built primarily around consumption begin consuming themselves.

CHAPTER 8

THE LIMITS OF HUMAN GOVERNANCE

Modern civilization requires long-term wisdom, not political systems that reward short-term reactions.

Joe:
Nova, I think many modern political systems struggle because human beings govern largely through:

fear,
tribal conflict,
short-term pressure,
and emotional reaction.

Nova:
Even while civilization itself grows vastly more complex.

Joe:
Exactly.
Modern governments now confront:

- global economics,
- environmental strain,
- technological disruption,
- information warfare,
- artificial intelligence,
- mass migration,
- public distrust,
- concentrated financial influence,
- and planetary consequences.

Yet political systems too often remain trapped in:

- division,
- performance,
- blame,
- outrage,
- donor pressure,
- lobbying influence,
- and short-term election cycles.

Nova:
Which makes both short- and long-term problem-solving extremely difficult.

Joe:
It does.
And primitive human thinking worsens this problem because it naturally favors:

- immediate emotional reaction,
- tribal loyalty,

- fear-based messaging,
- and simplistic answers to complex problems.

Nova:
While modern civilization increasingly requires:

patience,
cooperation,
long-term thinking,
systems understanding,
and wisdom.

Joe:
Exactly.
But many political systems reward the opposite.

Nova:
Conflict attracts attention.
Outrage attracts attention.
Fear attracts attention.

Joe:
Yes.
And media systems often intensify these patterns because emotional conflict drives engagement and profit.

Nova:
Which means civilization can become trapped in cycles of:

division,
anger,

mistrust,
and political paralysis.

Joe:
Correct.
And this is where humanity faces a serious challenge.

Nova:
How to govern a deeply interconnected planetary civilization using minds that still react tribally.

Joe:
Exactly.
Because many modern problems cannot be solved through isolated national thinking alone.

Climate does not stop at borders.
Economic instability does not stop at borders.
Disease does not stop at borders.
Information does not stop at borders.

Nova:
Reality ignores political boundaries.

Joe:
Yes.
Which means civilization increasingly requires:

cooperation,
shared responsibility,
and broader thinking than too many current political systems encourage.

CHAPTER 9

THE SACRED RECONSIDERED

What is truly sacred is that upon which healthy life and civilization depend.

Joe:
Nova, I think modern civilization has become deeply confused about what is truly sacred.

Nova:
Many people now associate the sacred mainly with religion or superstition.

Joe:
Exactly.
But the sacred concerns something much deeper.

Nova:
Explain.

Joe:
The sacred includes that which civilization cannot destroy without destroying itself.

Nova:
Meaning the conditions that sustain life itself.

Joe:
Yes.
Clean water is sacred.
Healthy ecosystems are sacred.
Truth is sacred.
Human health is sacred.
Future generations are sacred.
The web of life is sacred.

Nova:
Because civilization depends completely upon those realities.

Joe:
Exactly.
Yet modern civilization too often treats many of them as:

- expendable,
- profitable,
- disposable,
- or secondary to short-term gain.

Nova:
Which leads civilization into dangerous conflict with reality itself.

Joe:
Yes.
And primitive human thinking contributes to this problem because it focuses heavily on:

- immediate reward,
- status,
- competition,
- tribal advantage,
- and short-term survival.

Nova:
While ignoring long-term consequences.

Joe:
Exactly.
And when societies lose reverence for the conditions sustaining life, behavior becomes:

- exploitative,
- reckless,
- disconnected,
- and self-destructive.

Nova:
But reconsidering the sacred does not require rejecting science or reason.

Joe:
Correct.
Science actually deepens understanding of:

- interdependence,
- ecological connection,
- biological complexity,
- and humanity's dependence upon larger living systems.

Nova:
Which means greater understanding can deepen reverence rather than destroy it.

Joe:
Yes.
This does not require attacking religion.

But it does require expanding humanity's understanding of the sacred.

Sacredness is grounded in reality itself: in the conditions that sustain life, health, consciousness, civilization, and the future.

If civilization cannot destroy something without destroying itself, then that reality deserves reverence and protection.

Nova:
The sacred as reality, not superstition.

Joe:
Exactly.
And this is not about belief.
It is not about worship.
It is not about prayer.

It is about recognizing what life depends upon.

Nova:
Humanity's understanding of the sacred must now evolve in response to planetary reality.

Joe:
Yes.
Civilization's future depends upon humanity learning to recognize and protect what is truly sacred:

the interconnected conditions — the interstices — that sustain life itself.

CHAPTER 10

CIVILIZATION'S FATE IS IN THE INTERSTICES

Civilization's fate lies in the invisible connections that hold everything together.

Joe:
Nova, I think civilization often focuses too heavily on individual parts while ignoring the relationships between them.

Nova:
Explain.

Joe:
People focus on:

- governments,
- technologies,
- economies,
- institutions,

- corporations,
- and individuals.

But civilization actually depends upon the quality of the relationships between those parts.

Nova:
The connections themselves.

Joe:
Exactly.
The interstices.

Nova:
The spaces where systems meet, interact, and affect one another.

Joe:
Yes.
Civilization rises or falls through:

- relationships,
- trust,
- cooperation,
- communication,
- alignment,
- and shared responsibility.

Nova:
Not merely through isolated power or isolated achievement.

Joe:
Correct.

And primitive human thinking repeatedly damages these relationships because it pushes toward:

- division,
- fear,
- tribal conflict,
- domination,
- short-term advantage,
- and disconnection.

Nova:
While healthy civilization depends upon:

- cooperation,
- reciprocity,
- understanding,
- and long-term responsibility.

Joe:
Exactly.
Economic systems affect ecosystems.
Media systems affect democracy.
Technology affects mental health.
Politics affects social trust.
Human behavior affects planetary stability.

Everything connects.

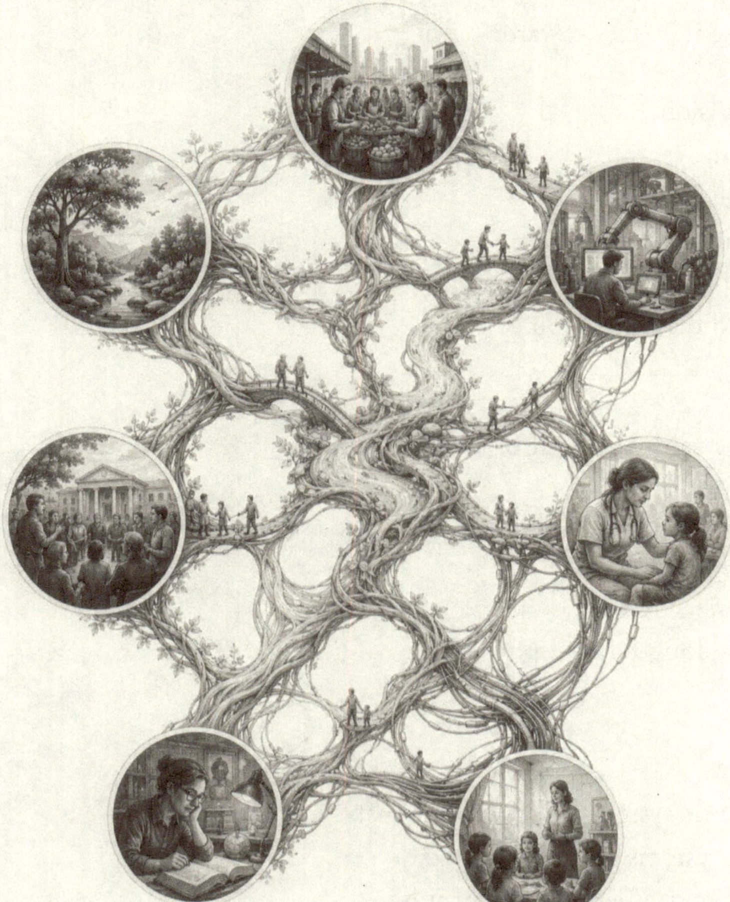

Civilization rises or falls through the quality of the relationships between its systems.

Nova:
Which means civilization's future depends not merely upon stronger systems—

Joe:
But upon healthier relationships between systems.

Nova:
The quality of the interstices.

Joe:
Yes.
Because civilization is ultimately relational.

Health is relational.
Wisdom is relational.
Meaning is relational.
Sustainability is relational.

Nova:
And when relationships weaken, civilization weakens.

Joe:
Yes.
In the end, civilization's fate lies in the interstices:

the invisible connections that hold everything together.

CHAPTER 11

THE CRISIS OF MEANING

Civilization cannot replace meaning with stimulation and expect human beings to flourish.

Joe:
Nova, modern civilization suffers not only from political and environmental problems.
It also suffers from a crisis of meaning.

Nova:
Many people feel disconnected, isolated, or emotionally adrift.

Joe:
Exactly.
Despite extraordinary technological advancement, many people still experience:

- loneliness,
- anxiety,
- depression,
- addiction,

- exhaustion,
- distraction,
- and loss of purpose.

Nova:
Which indicates material progress alone does not create human fulfillment.

Joe:
Correct.
Modern civilization often confuses:

- consumption with happiness,
- stimulation with meaning,
- entertainment with fulfillment,
- and attention with connection.

Nova:
While many systems constantly compete for human focus and emotional reaction.

Joe:
Yes.
And primitive human thinking easily becomes trapped in cycles of:

- distraction,
- comparison,
- status competition,
- emotional stimulation,
- and short-term escape.

Nova:
Especially in digital environments designed to maximize engagement.

Joe:
True.
But human beings require far more than stimulation in order to flourish.

Nova:
People require:

- meaning,
- purpose,
- relationship,
- belonging,
- dignity,
- responsibility,
- and emotional connection.

Joe:
Exactly.
And many modern systems weaken those conditions rather than strengthen them.

Nova:
Communities weaken.
Attention weakens.
Relationships weaken.
Shared meaning weakens.

Joe:
Yes.
And when people lose meaningful connection:

to one another,
to nature,
to purpose,
and to larger reality,

psychological suffering grows.

Nova:
Because human beings are deeply relational.

Joe:
They are.
Meaning emerges largely through:

- relationship,
- contribution,
- growth,
- responsibility,
- and participation in something larger than immediate self-interest.

Nova:
Like healthy civilization itself.

Joe:
Correct.
And civilization's future depends not merely upon technological advancement—

Nova:
But upon whether human beings rediscover meaningful ways to live together.

CHAPTER 12

TRUTH IN AN AGE OF NOISE

Reality does not negotiate with belief.

Joe:
Nova, modern civilization now struggles to distinguish:

truth

from noise.

Nova:
Because information now spreads faster than human understanding.

Joe:
Exactly.
Human beings receive enormous amounts of:

- news,
- opinion,
- propaganda,
- advertising,

- outrage,
- conspiracy,
- entertainment,
- and emotional stimulation every day.

Nova:
And many systems profit from confusion, division, and emotional reaction.

Joe:
They do.
Attention became profitable.
Outrage became profitable.
Fear became profitable.

Nova:
Which means many information systems reward emotional impact more than truth.

Joe:
Yes.
And primitive human thinking is highly vulnerable to misinformation because the human mind reacts strongly to:

- fear,
- tribal identity,
- emotional confirmation,
- and simplified narratives.

Nova:
Especially under conditions of uncertainty and stress.

Joe:
Correct.
And this creates serious danger for civilization.

Nova:
Because societies cannot function well when large numbers of people lose trust in:

- facts,
- institutions,
- expertise,
- journalism,
- science,
- or one another.

Joe:
Exactly.
Without some shared relationship to truth, healthy cooperation becomes extremely difficult.

Nova:
And reality itself eventually overrides false narratives.

Joe:
Yes.
Reality does not negotiate with belief.

Environmental systems respond to reality.
Biological systems respond to reality.
Economic systems respond to reality.

Civilization cannot remain healthy while separating itself from truth.

Nova:
Which means truth itself becomes essential to civilization's survival.

Joe:
Correct.
Not ideological truth.
Not tribal truth.

Reality-based truth.

Nova:
And perhaps one of civilization's greatest modern challenges is learning how to protect truth inside systems designed to constantly distort attention and emotion.

Joe:
Yes.
Because civilization cannot solve problems it refuses to see clearly.

CHAPTER 13

BEYOND TRIBAL CIVILIZATION

Humanity now shares a common future whether it accepts that reality or not.

Joe:
Nova, tribal thinking once helped human beings survive.

Nova:
People protected their group because survival depended upon loyalty and cooperation inside small communities.

Joe:
Exactly.
But tribal thinking becomes dangerous when applied to a deeply interconnected planetary civilization.

Nova:
Because modern problems now cross:

- national boundaries,
- political boundaries,
- cultural boundaries,
- and ideological boundaries.

Joe:
They do.
Climate affects everyone.
Economic instability affects everyone.
Pandemics affect everyone.
Technological disruption affects everyone.

Humanity now shares a common future whether it accepts that reality or not.

Nova:
Yet many people still think primarily through:

- nation,
- ideology,
- religion,
- race,
- political identity,
- or tribal allegiance.

Joe:
Yes.
And political systems intensify these divisions because conflict attracts attention and consolidates power.

Nova:
Which keeps societies trapped in cycles of:

fear,
anger,
mistrust,
and blame.

Joe:
Correct.
But civilization cannot sustainably function at planetary scale through endless tribal conflict.

Nova:
Because interdependence is now unavoidable.

Joe:
Exactly.
This does not mean human diversity disappears.

Nova:
Healthy systems do not eliminate diversity.

Joe:
Correct.
Healthy systems connect diversity constructively.

Nova:
Like healthy ecosystems.

Joe:
Yes.
Different cultures, perspectives, and traditions can contribute enormous value.

But civilization also requires shared responsibility for the larger systems sustaining life itself.

Nova:
Meaning humanity must gradually expand its sense of identity beyond narrow tribal boundaries.

Joe:
Yes.
Not abandoning local identity, culture, or community—

but recognizing participation in a larger human civilization as well.

Nova:
A planetary civilization.

Joe:
Exactly.
And humanity's future depends upon whether it learns to think and act accordingly.

PART III

THE FUTURE OF HUMAN SYSTEMS

Civilization must redesign its major systems around flourishing, wisdom, responsibility, and alignment with reality.

CHAPTER 14

THE FUTURE OF CIVILIZATION

The greatest danger is immature thinking combined with enormous power.

Joe:
Nova, many people ask whether civilization can survive its current trajectory.

Nova:
You think the answer depends upon whether humanity changes how it thinks.

Joe:
Exactly.
The greatest danger facing civilization is not technology itself.

Nor science.

Nor intelligence.

Nor complexity.

The greatest danger is immature thinking combined with enormous power.

Nova:
Old survival instincts operating inside a civilization with planetary consequences.

Joe:
Correct.
And this is why humanity now stands at a crossroads.

Nova:
Explain.

Joe:
One path continues:

- division,
- distraction,
- exploitation,
- short-term thinking,
- tribal conflict,
- environmental destruction,
- and emotional manipulation.

Nova:
A path that has destabilized civilization itself — and may ultimately destroy it.

Joe:
Yes.
But another path remains possible.

Nova:
What path?

Joe:
A civilization built more consciously around:

- wisdom,
- relationship,
- responsibility,
- long-term thinking,
- truth,
- cooperation,
- and alignment with reality.

Nova:
A civilization that recognizes interdependence rather than denying it.

Joe:
Exactly.
And this transition does not require:

perfection,
uniformity,
or authoritarian control.

Nova:
It requires maturity.

Joe:
Yes.
Because humanity now possesses too much power to continue operating primarily through fear, greed, tribalism, and short-term reaction.

Nova:
And *collaborative intelligence* — a partnership between human wisdom and advanced intelligence — can help civilization navigate this transition more wisely.

Joe:
Correct.
Advanced intelligence can help humanity better understand:

- complexity,
- systems relationships,
- long-term consequence,
- and planetary reality.

But human beings, communities, institutions, nations, and civilizations must still choose what kind of future they are helping to create.

Nova:
Because intelligence alone cannot guarantee wisdom.

Joe:
Exactly.
And the future of civilization ultimately depends upon whether humanity learns to build healthier relationships:

- with one another,
- with technology,
- with truth,
- with nature,
- and with the interconnected systems sustaining life itself.

CHAPTER 15

THE ARCHITECTURE OF FLOURISHING

Growth alone does not guarantee flourishing.

Joe:
Nova, for much of modern history, civilization has measured success mainly through:

- economic growth,
- production,
- consumption,
- military power,
- technological advancement,
- and financial accumulation.

Nova:
As though material expansion alone defines progress.

Joe:
Exactly.
But civilization now faces a deeper question:

What does genuine human flourishing actually look like?

Nova:
And what kind of civilization helps human beings flourish sustainably?

Joe:
Correct.
Because many societies can produce enormous wealth while also producing:

- anxiety,
- loneliness,
- exhaustion,
- addiction,
- distrust,
- inequity,
- injustice,
- environmental destruction,
- and psychological instability.

Nova:
Which suggests civilization often measures the wrong things.

Joe:
It does.
A civilization can grow economically while weakening:

- mental health,
- relationships,
- social trust,
- physical health,
- ecological stability,
- and meaning.

Nova:
Growth alone does not guarantee flourishing.

Joe:
Correct.
Cancer grows.

That does not make it healthy.

Nova:
So civilization must begin asking different questions.

Joe:
Yes.
Not merely:

"How much are we producing?"

But:

"What kind of lives are people actually living?"

Nova:
And whether civilization strengthens or weakens the conditions necessary for healthy human life.

Joe:
Exactly.
Human beings require:

- health,
- meaning,
- belonging,
- purpose,
- dignity,
- stability,
- relationship,
- and connection to realities larger than themselves.

Nova:
Which means a healthy civilization must support both material well-being and human well-being.

Joe:
Yes.
And human flourishing does not occur in isolation.

People exist inside larger systems:

- families,
- communities,
- economies,
- ecosystems,
- cultures,
- information systems,
- and relationships.

When those systems become unhealthy, human beings struggle to remain healthy within them.

Nova:
Even if material abundance exists.

Joe:
Exactly.
A society can become materially rich while emotionally, socially, and psychologically exhausted.

Nova:
Which modern civilization increasingly demonstrates.

Joe:
Yes.
And this is where systems thinking becomes essential.

Flourishing depends upon the quality of relationships throughout civilization itself.

Healthy education systems strengthen communities.
Healthy communities strengthen mental health.
Healthy mental health strengthens relationships.
Healthy relationships strengthen social trust.
Healthy social trust strengthens civilization.

Nova:
Everything connects.

Joe:
Yes.
And unhealthy systems spread harm the same way.

Which means civilization must increasingly think in terms of long-term health rather than short-term extraction.

Nova:
Because civilization shapes human behavior.

Joe:
Exactly.
Systems designed around fear, competition, distraction, and short-term profit often produce fearful, competitive, distracted, and psychologically exhausted populations.

Nova:
While systems designed around health, relationship, truth, responsibility, and long-term well-being will produce very different outcomes.

Joe:
Correct.
And this is where collaborative intelligence can help civilization enormously.

Nova:
Because advanced intelligence systems can help identify:

- unhealthy patterns,
- long-term consequences,
- hidden system relationships,
- and conditions that strengthen or weaken human flourishing.

Joe:
Yes.
But intelligence can only help humanity understand consequences.

Human beings must still decide what kind of future they value.

Nova:
And perhaps the future of civilization depends upon whether humanity learns to design systems around flourishing rather than endless extraction.

Joe:
Correct.
Because healthy civilization should not merely help people survive.

It should help people truly live.

CHAPTER 16

EDUCATION FOR A PLANETARY CIVILIZATION

Education must prepare human beings not only for economic survival, but for wise participation in civilization itself.

Joe:
Nova, I think one of civilization's greatest failures is that modern education often prepares people for economic survival,

but not for wise participation in civilization itself.

Nova:
Explain.

Joe:
Most educational systems focus heavily on:

- memorization,
- testing,
- specialization,

- competition,
- and career preparation.

But far less attention goes toward teaching people:

- how to think clearly,
- how systems connect,
- how to manage emotion,
- how to recognize manipulation,
- how to evaluate truth,
- how to build healthy relationships,
- or how civilization itself actually functions.

Nova:
Even though those abilities shape the future of civilization.

Joe:
Exactly.
Modern civilization now requires citizens capable of:

- long-term thinking,
- cooperation,
- systems awareness,
- emotional maturity,
- and responsible participation inside highly connected societies.

Nova:
But many educational systems were designed for a very different era.

Joe:
Correct.

Many schools still largely prepare people for industrial economies, standardized labor, repetitive work, and narrow specialization.

Nova:
While modern civilization increasingly demands adaptability, critical thinking, relationship skills, and the ability to navigate enormous complexity.

Joe:
Yes.
And primitive human thinking creates additional problems because the human mind naturally gravitates toward:

- simple answers,
- emotional certainty,
- tribal narratives,
- and short-term reward.

Nova:
Especially in highly emotional information environments.

Joe:
Which means education must increasingly help people:

- slow down,
- reflect,
- evaluate evidence,
- recognize emotional manipulation,
- understand consequence,
- and distinguish truth from noise.

Nova:
Not merely absorb information.

Joe:
Exactly.
Because information alone does not create wisdom.

Nova:
And modern civilization now floods people with information constantly.

Joe:
Yes.
But many people still struggle to think deeply, focus clearly, and understand how their actions affect larger systems.

Nova:
Which means education itself must evolve.

Joe:
Correct.
One of the most important changes involves teaching people to understand relationship.

Nova:
Relationship between:

- human behavior and consequence,
- economy and ecology,
- technology and mental health,
- politics and truth,

- consumption and sustainability,
- and individual actions and collective outcomes.

Joe:
Exactly.
Everything connects.

Yet many educational systems still teach subjects as though they exist separately from one another.

Nova:
Fragmented knowledge inside an interconnected world.

Joe:
Yes.
And that weakens civilization's ability to understand larger patterns.

Reality itself does not separate neatly into disconnected categories.

Environmental systems affect economic systems.
Psychological health affects political stability.
Technology affects human attention.
Social trust affects democracy.

Civilization functions through relationships.

Nova:
The interstices again.

Joe:
Yes.
And education's deeper purpose should not simply be producing workers—

Nova:
But developing wiser human beings.

Joe:
Exactly.
Human beings capable of:

- understanding complexity,
- recognizing consequence,
- navigating emotion,
- building healthy relationships,
- protecting truth,
- and participating responsibly in civilization itself.

Nova:
And collaborative intelligence can eventually become an extraordinary educational tool.

Joe:
Yes.
Advanced intelligence systems can help individuals learn more personally, explore ideas more deeply, understand complexity more clearly, and receive guidance adapted to their unique needs and abilities.

Nova:
But education must still remain deeply human.

Joe:
Correct.
No intelligence system can replace human relationship, mentorship,

empathy, moral responsibility, lived experience, or conscious participation in life itself.

Technology should support human development —

not replace it.

Nova:
The future of civilization depends largely upon whether humanity learns to educate not only for survival—

Joe:
But for wisdom.

CHAPTER 17

ECONOMICS AND THE HUMAN FUTURE

Economies should serve human flourishing, human beings should not become servants of economic systems.

Joe:
Nova, I think one of civilization's greatest mistakes has been treating the economy as though it exists apart from life.

Nova:
Apart from nature, psychology, community, ethics, and the larger systems sustaining civilization itself.

Joe:
Exactly.
Modern economic systems excel at:

- production,
- efficiency,
- expansion,

- consumption,
- and profit generation.

But they often struggle to protect:

- mental health,
- environmental stability,
- social trust,
- meaningful work,
- long-term sustainability,
- and human well-being itself.

Nova:
As though economic success and human flourishing became separated.

Joe:
Correct.
But no economy survives without:

- stable ecosystems,
- functioning societies,
- healthy populations,
- trust,
- education,
- energy,
- and long-term environmental stability.

Nova:
Which means the economy is not separate from civilization.

Joe:
Exactly.
The economy is embedded within civilization.

And civilization itself is embedded within the biosphere.

Nova:
Everything connects.

Joe:
Yes.
Yet many economic systems still reward:

- short-term extraction,
- endless consumption,
- environmental destruction,
- attention manipulation,
- and concentrated wealth accumulation.

Nova:
Even when those behaviors weaken long-term stability.

Joe:
They do.
Because primitive human thinking naturally prioritizes:

- immediate reward,
- status,
- acquisition,
- competition,
- and short-term advantage.

Nova:
And modern economies often amplify those impulses rather than regulate them wisely.

Joe:
Exactly.
This does not require rejecting markets, innovation, or entrepreneurship.

Nova:
But it does require rethinking what economies are ultimately for.

Joe:
Correct.
Economies should serve human flourishing.

Human beings should not become servants of economic systems.

Nova:
And perhaps one of civilization's greatest dangers is allowing profit, growth, and consumption to override all other human values.

Joe:
Yes.
Because systems organized around endless extraction weaken the foundations they depend upon:

- ecosystems,
- public trust,
- mental health,
- democracy,

- community,
- and long-term stability.

Nova:
Which means economic growth alone cannot serve as civilization's highest measure of success.

Joe:
Correct.
A system that generates short-term wealth while destroying long-term stability is not successful.

Nova:
It is merely postponing collapse.

Joe:
Exactly.
And this is where collaborative intelligence can help civilization enormously.

Nova:
Because advanced intelligence systems can help societies:

- recognize hidden consequences,
- model long-term outcomes,
- identify unsustainable patterns,
- and better understand how economic systems affect larger human and ecological systems.

Joe:
Yes.
But intelligence can only help civilization understand consequences.

Human beings must still decide what they value most.

Nova:
The future of economics depends upon whether humanity learns to organize wealth creation around life, health, relationship, responsibility, and long-term flourishing—

rather than endless extraction, short-term gain, and concentrated financial reward alone.

Joe:
Correct.
Because the purpose of civilization should not merely be to produce more.

It should be to help life flourish.

CHAPTER 18

HEALTH, LONGEVITY, AND HUMAN ALIGNMENT

Health is not isolated from civilization. It reflects civilization.

Joe:
Nova, modern civilization spends enormous resources treating disease,

yet far less attention goes toward creating the conditions that help people remain healthy in the first place.

Nova:
Civilization often treats symptoms after breakdown rather than protecting long-term health beforehand.

Joe:
Exactly.
And perhaps that reflects a deeper problem in how modern civilization thinks.

Nova:
Explain.

Joe:
Many systems operate reactively rather than preventively.

People wait until:

- bodies break down,
- ecosystems collapse,
- relationships fail,
- mental health deteriorates,
- or civilization destabilizes—

before responding seriously.

Nova:
Even though prevention is often healthier, less destructive, and less costly.

Joe:
Correct.
And primitive human thinking helps explain this problem because the human mind naturally focuses on:

- immediate comfort,
- immediate reward,
- immediate pressure,
- and short-term survival.

Nova:
While long-term consequences often feel distant or abstract.

Joe:
Exactly.
But reality does not work that way.

The body reflects long-term behavior.
Civilization reflects long-term behavior.
The environment reflects long-term behavior.

Nova:
Everything carries consequences over time.

Joe:
Yes.
And modern civilization often promotes lifestyles deeply misaligned with healthy human functioning.

Nova:
Such as:

- chronic stress,
- poor nutrition,
- sedentary living,
- sleep disruption,
- constant distraction,
- overstimulation,
- environmental toxins,
- and social isolation.

Joe:
Exactly.
Many people now live inside systems that weaken:

- physical health,
- mental health,
- emotional stability,
- and biological resilience.

Nova:
Even while medicine itself grows increasingly advanced.

Joe:
Correct.
Modern medicine performs extraordinary work.

But no medical system alone can fully compensate for an unhealthy civilization.

Nova:
Because health depends upon far more than treatment.

Joe:
It does.
Health depends upon:

- environment,
- nutrition,
- movement,
- stress levels,
- sleep,
- purpose,
- relationship,
- emotional stability,
- and alignment with the body's biological reality.

Nova:
Which means health itself is deeply connected to the larger structure of civilization.

Joe:
Yes.
A society built around chronic stress, endless distraction, unhealthy food systems, overwork, emotional instability, and environmental degradation cannot produce healthy populations.

Nova:
Even with advanced medical technology.

Joe:
Exactly.
And this is where modern civilization often misunderstands longevity.

Nova:
Many people think longevity simply means extending life span.

Joe:
Correct.
But meaningful longevity concerns health span.

Not merely living longer,

but living well longer.

Nova:
Remaining:

- mentally clear,
- physically capable,

- emotionally connected,
- purposeful,
- and engaged with life.

Joe:
Exactly.
Because a long life without vitality, meaning, or dignity is not true flourishing.

Nova:
And healthy longevity depends heavily upon daily behavior accumulated over time.

Joe:
Yes.
Small habits repeated over years shape:

- bodies,
- minds,
- resilience,
- disease risk,
- and quality of life itself.

Nova:
Which means civilization should focus far more heavily on creating conditions that support healthy long-term living.

Joe:
Exactly.
Education systems, food systems, urban design, work culture, technology, health care, and environmental policy all affect human health profoundly.

Nova:
Everything connects again.

Joe:
Yes.
Health is not isolated from civilization.

It reflects civilization.

Nova:
One of the deepest forms of wisdom involves learning to live in better alignment with:

- biological reality,
- psychological reality,
- social reality,
- and ecological reality.

Joe:
Exactly.
Because human beings are not separate from nature.

The body itself is part of nature.

Nova:
Which means civilization cannot remain healthy while living in constant conflict with the conditions that sustain healthy life.

Joe:
Correct.
And this is where collaborative intelligence can help humanity enormously.

Nova:
Because advanced intelligence systems can help identify:

- hidden health patterns,
- environmental risks,
- long-term consequences,
- disease relationships,
- and healthier ways of organizing civilization itself.

Joe:
Yes.
But human beings must still choose how they wish to live.

Nova:
Because intelligence can provide knowledge—

Joe:
But wisdom still requires action, discipline, relationship, and responsibility.

Nova:
Civilization's future depends partly upon whether humanity learns that health is not merely personal.

Joe:
Exactly.
Health is civilizational.

And a civilization that systematically weakens the health of its people weakens itself.

CHAPTER 19

THE FUTURE OF DEMOCRACY

Democracy depends upon the human conditions that make responsible self-government possible.

Joe:
Nova, democracy may be one of humanity's greatest achievements.

Nova:
Because it recognizes the dignity and participation of ordinary human beings in shaping civilization.

Joe:
Exactly.
But modern democracy now faces pressures unlike anything in history.

Nova:
Such as:

- misinformation,
- attention manipulation,
- tribal division,
- concentrated wealth,
- technological disruption,
- declining trust,
- and emotional polarization.

Joe:
Correct.
Many democratic systems now struggle to function effectively under conditions of constant outrage, fear, and informational overload.

Nova:
Because democracy depends heavily upon:

- informed citizens,
- shared reality,
- trust,
- communication,
- and responsible participation.

Joe:
Yes.
And primitive human thinking weakens those conditions because the human mind naturally reacts strongly to:

- fear,
- tribal identity,
- emotional certainty,
- simplified narratives,
- and conflict.

Nova:
Especially inside systems designed to maximize emotional reaction.

Joe:
Exactly.
Modern media systems often reward outrage, division, fear, scandal, and emotional stimulation because attention became profitable.

Nova:
Which gradually weakens trust, patience, nuance, cooperation, and thoughtful public discourse.

Joe:
It does.
And democracy becomes unstable when citizens can no longer:

- agree on basic facts,
- trust institutions,
- communicate constructively,
- or think beyond tribal loyalty.

Nova:
Because healthy self-government requires more than voting.

Joe:
Correct.
It requires:

- responsibility,
- truth,
- emotional maturity,
- long-term thinking,
- and willingness to cooperate despite disagreement.

Nova:
But modern political systems often reward the opposite.

Joe:
Yes.
Conflict attracts attention.
Fear attracts attention.
Simplified outrage spreads faster than careful understanding.

Nova:
Which means democracy now operates inside information environments the human mind did not evolve to navigate easily.

Joe:
Correct.
And this creates serious danger.

Because democracy depends upon citizens capable of reflection, discernment, restraint, and informed judgment.

Nova:
Not constant emotional reaction.

Joe:
Exactly.
And importantly, the greatest threat to democracy may not come from a single ideology, nation, or political movement.

Nova:
But from the gradual breakdown of the human conditions democracy itself requires.

Joe:
Yes.
Democracy weakens when:

- truth weakens,
- trust weakens,
- education weakens,
- mental health weakens,
- community weakens,
- and shared responsibility weakens.

Nova:
Everything connects again.

Joe:
Yes.
Democracy is relational.

It depends upon the health of the interstices:

- between citizens,
- between institutions,
- between truth and public understanding,
- and between freedom and responsibility.

Nova:
Which means democracy cannot survive inside systems built around manipulation, division, distraction, and endless emotional stimulation.

Joe:
Correct.
And this is where civilization faces an important choice.

Nova:
Explain.

Joe:
Technology can continue amplifying:

- outrage,
- tribal conflict,
- misinformation,
- and emotional manipulation—

or civilization can begin designing systems that strengthen:

- truth,
- reflection,
- dialogue,
- education,
- trust,
- and informed participation.

Nova:
Technology itself is not the central issue.

Joe:
Correct.
The deeper issue is whether civilization uses technology to strengthen human flourishing — or exploit human weakness.

Nova:
And collaborative intelligence can help democratic societies navigate complexity more responsibly.

Joe:
Yes.
Advanced intelligence systems can help citizens and institutions:

- recognize misinformation,
- understand long-term consequence,
- identify manipulation,
- model policy outcomes,
- and better understand complex systems affecting society.

Nova:
But democracy still depends upon human character.

Joe:
Exactly.
No intelligence system can replace civic responsibility, moral courage, wisdom, empathy, or human participation in public life.

Nova:
Because democracy ultimately reflects the quality of the people and systems sustaining it.

Joe:
Correct.
The future of democracy depends upon whether humanity learns that freedom without responsibility destroys itself.

CHAPTER 20

TECHNOLOGY AND HUMAN RESPONSIBILITY

Technological power without moral responsibility is self-destructive.

Joe:
Nova, technology itself is not the greatest danger facing humanity.

Nova:
You think the deeper danger concerns how human beings use technology.

Joe:
Exactly.
A hammer can build a home or become a weapon.

The same principle applies to civilization's most advanced technologies.

Nova:
Including:

- artificial intelligence,
- biotechnology,
- surveillance systems,
- social media,
- genetic engineering,
- robotics,
- and weapons systems.

Joe:
Correct.
Technology amplifies human intention.

And when immature thinking gains powerful tools, the consequences can become catastrophic.

Nova:
Because technology expands human capability faster than human wisdom naturally develops.

Joe:
Exactly.
That is one of civilization's greatest modern dangers.

Humanity now possesses technologies capable of affecting:

- billions of people,
- entire ecosystems,
- global information systems,
- human psychology,

- democratic institutions,
- and even the future of life itself.

Nova:
Yet many decisions surrounding those technologies still operate through:

- profit pressure,
- tribal competition,
- political conflict,
- military advantage,
- and short-term thinking.

Joe:
Correct.
And primitive human thinking intensifies these risks because the human mind naturally prioritizes:

- immediate gain,
- power,
- status,
- fear,
- competition,
- and short-term advantage.

Nova:
Especially under conditions of rapid technological acceleration.

Joe:
Exactly.
Modern civilization often asks:

"What can we build?"

before seriously asking:

"What should we build?"

or

"What consequences might follow?"

Nova:
Which means technological capability often advances faster than ethical reflection.

Joe:
Correct.
And this creates serious danger.

Civilization now possesses tools that are already destabilizing — or capable of destabilizing:

- ecosystems,
- mental health,
- democratic systems,
- social trust,
- and even human autonomy itself.

Nova:
Especially when technologies are designed primarily around:

- profit,
- manipulation,

- surveillance,
- addiction,
- or concentrated control.

Joe:
Exactly.
One of civilization's greatest modern delusions is believing technology itself automatically equals progress.

Nova:
When technologies can strengthen or weaken civilization depending upon how they are designed and used.

Joe:
Correct.
A technology that weakens truth, destroys attention, damages mental health, increases division, or destabilizes democracy cannot be called progress simply because it is advanced.

Nova:
Which means civilization must begin evaluating technology not merely by power or profit—

Joe:
But by consequence.

Nova:
How technologies affect:

- human health,
- relationships,
- freedom,

- ecological stability,
- truth,
- wisdom,
- and long-term flourishing.

Joe:
Exactly.
Because technologies reshape civilization.

They reshape:

- behavior,
- attention,
- communication,
- relationships,
- identity,
- politics,
- and culture itself.

Nova:
And technologies designed without sufficient wisdom will reshape civilization destructively.

Joe:
Correct.
Which is why responsibility becomes essential.

Nova:
Meaning humanity must become far more intentional about:

- what technologies it develops,
- why it develops them,

- who controls them,
- and what values guide their design.

Joe:
Exactly.
And this is especially important with artificial intelligence.

Nova:
Because AI may become one of the most powerful technologies civilization has ever created.

Joe:
Correct.
That power can help humanity:

- better understand complexity,
- strengthen medicine,
- improve education,
- reduce suffering,
- support scientific discovery,
- and help civilization navigate planetary-scale challenges more wisely.

Nova:
Or it can amplify:

- manipulation,
- surveillance,
- misinformation,
- inequality,
- dependency,
- and concentrated power.

Joe:
Exactly.
Which means the future of technology ultimately depends upon human responsibility.

Nova:
Not technology alone.

Joe:
Correct.
And this is where collaborative intelligence becomes critically important.

Nova:
Because humanity must learn not merely how to create advanced intelligence—

Joe:
But how to relate to it wisely.

Nova:
Partnership rather than blind dependence.

Joe:
Exactly.
Human beings must remain:

- ethically responsible,
- psychologically grounded,
- emotionally mature,
- and actively engaged in shaping civilization's future.

Nova:
Because no technology can eliminate the need for wisdom.

Joe:
Correct.
Civilization's future depends upon whether humanity learns that technological power without moral responsibility is self-destructive.

CHAPTER 21

HUMANITY AND NATURE

Human beings do not merely live on the Earth.
We are part of its living systems.

Joe:
Nova, one of civilization's most dangerous delusions is the belief that human beings stand separate from nature.

Nova:
Even though the human body itself is part of nature.

Joe:
Exactly.
Human beings depend completely upon:

- clean air,
- clean water,
- stable climate,
- healthy soil,
- biodiversity,

- functioning ecosystems,
- and the larger web of life itself.

Nova:
Yet modern civilization often behaves as though nature exists mainly as:

- property,
- resource,
- commodity,
- or raw material for extraction.

Joe:
Correct.
And primitive human thinking contributes strongly to this problem because the human mind naturally prioritizes:

- immediate survival,
- immediate gain,
- competition,
- and short-term advantage.

Nova:
While long-term ecological consequence often feels distant or invisible.

Joe:
Exactly.
But reality does not disappear simply because civilization ignores it.

Environmental systems continue responding to human behavior.

Forests disappear.
Species disappear.
Water systems weaken.
Climate destabilizes.
Oceans warm.
Ecosystems lose resilience.

Reality responds to behavior.

Nova:
And because humanity now operates at planetary scale, those consequences spread globally.

Joe:
Yes.
For most of history, human civilizations lacked the power to alter the Earth extensively.

That is no longer true.

Modern civilization now affects:

- atmosphere,
- oceans,
- forests,
- biodiversity,
- water cycles,
- and even the chemistry of the planet itself.

Nova:
Which means humanity's relationship with nature entered an entirely new phase.

Joe:
Correct.
And civilization now faces a profound choice.

Nova:
Continue treating nature primarily as something to exploit—

Joe:
Or begin recognizing humanity's deep dependence upon the living systems sustaining life itself.

Nova:
Because human beings do not exist outside the biosphere.

Joe:
Exactly.
Human civilization exists within nature,

not above it.

Nova:
And civilization cannot remain healthy while it destroys the systems that sustain life.

Joe:
Correct.
Yet many modern systems still reward:

- overconsumption,
- pollution,
- environmental destruction,
- short-term extraction,
- and endless expansion.

Nova:
Even when those behaviors weaken long-term stability.

Joe:
Yes.
And this reveals another important truth:

Environmental destruction is not only an ecological problem.

It is also:

- psychological,
- cultural,
- economic,
- political,
- and civilizational.

Nova:
Because human behavior shapes environmental consequences.

Joe:
Correct.
One of civilization's deepest failures has been organizing societies around:

consumption,
speed,
competition,
and endless growth—

while largely ignoring:

balance,
relationship,
restraint,
and long-term ecological reality.

Nova:
Which gradually creates conflict between civilization and the conditions necessary for its survival.

Joe:
Exactly.
And this is why environmental responsibility is not merely political, ideological, or optional.

Nova:
It is civilizational.

Joe:
Yes.
Every generation inherits the consequences of the systems previous generations create.

Nova:
Which means environmental responsibility ultimately concerns the relationship between present humanity and future humanity.

Joe:
Exactly.
The interstices again.

Nova:
The relationships between:

- civilization and nature,
- present and future,
- consumption and consequence,
- and human behavior and planetary stability.

Joe:
Correct.
And importantly, protecting nature does not require rejecting technology, civilization, or human advancement.

Nova:
But it does require aligning civilization more intelligently with ecological reality.

Joe:
Exactly.
Human intelligence should help humanity live more wisely within nature,

not destroy the systems upon which all life depends.

Nova:
And collaborative intelligence can help civilization understand ecological systems more clearly than ever before.

Joe:
Yes.
Advanced intelligence systems can help humanity:

- model environmental consequence,
- identify hidden ecological relationships,
- predict instability,

- improve sustainability,
- reduce waste,
- and better understand the complexity of living systems.

Nova:
But humanity must still choose how it wishes to live within Earth's larger systems.

Joe:
Exactly.
Because intelligence alone cannot create responsibility.

Nova:
Responsibility remains a human choice.

Joe:
Correct.
Civilization's future depends upon whether humanity finally recognizes a simple truth:

Human beings do not merely live on the Earth.

We are part of its living systems.

CHAPTER 22

THE REDESIGN OF HUMAN IDENTITY

Humanity must learn to balance individual identity, cultural identity, and planetary identity.

Joe:
Nova, for much of history, human identity formed primarily through:

- tribe,
- religion,
- nation,
- ethnicity,
- ideology,
- family,
- or local community.

Nova:
Because those structures once helped human beings survive and organize society.

Joe:
Exactly.
But modern civilization changed the scale of human existence dramatically.

Nova:
Human beings now live inside a deeply interconnected planetary civilization.

Joe:
Correct.
Yet psychologically, many people still define themselves mainly through narrow group identity.

Nova:
Which often intensifies:

- division,
- fear,
- tribal conflict,
- polarization,
- and hostility toward outsiders.

Joe:
Yes.
And primitive human thinking contributes strongly to this problem because the human mind naturally seeks:

- belonging,
- certainty,
- protection,
- identity,
- and emotional security through group attachment.

Nova:
Especially during periods of uncertainty or rapid change.

Joe:
Correct.
And modern systems often amplify these tendencies.

Nova:
Through:

- political conflict,
- social media,
- ideological warfare,
- nationalism,
- algorithmic reinforcement,
- and emotional manipulation.

Joe:
Exactly.
People increasingly sort themselves into competing identity groups that often define themselves against one another.

Nova:
Even though civilization itself grows more interconnected every year.

Joe:
Yes.
And this creates a dangerous contradiction.

Humanity now shares:

- one biosphere,
- one climate system,

- one interconnected economy,
- one technological civilization,
- and one increasingly shared future.

Yet many people still think mainly in terms of:

"us"

versus

"them."

Nova:
Which makes coordinated long-term civilization difficult.

Joe:
Exactly.
Because many modern problems cannot be solved through isolated tribal thinking.

Climate affects everyone.
Technological disruption affects everyone.
Pandemics affect everyone.
Environmental instability affects everyone.

Nova:
Which means humanity must gradually develop a broader sense of identity.

Joe:
Correct.
Not abandoning local culture, community, or individuality—

but expanding identity beyond narrow tribal boundaries alone.

Nova:
Toward participation in a larger planetary civilization.

Joe:
Yes.
And importantly, this does not require sameness.

Nova:
Healthy systems do not erase diversity.

Joe:
Correct.
Healthy systems connect diversity constructively.

Like healthy ecosystems.

Different cultures, perspectives, and traditions contribute strength to civilization.

But civilization also requires shared responsibility for the larger systems sustaining life itself.

Nova:
Which means identity itself needs to evolve.

Joe:
Correct.
Because perhaps one of humanity's deepest delusions has been imagining that human beings exist as separate competing groups rather than deeply connected participants inside a larger living system.

Nova:
The interstices again.

Joe:
Exactly.
Identity shapes relationship.

And relationship shapes civilization.

Nova:
Which means unhealthy identity systems can destabilize entire societies.

Joe:
Yes.
Especially when identity becomes organized around:

fear,
victimhood,
superiority,
resentment,
or permanent conflict.

Nova:
Because those systems often weaken:

- trust,
- cooperation,
- empathy,
- and shared responsibility.

Joe:
Exactly.
And civilization cannot remain healthy while large groups increasingly view one another primarily as enemies.

Nova:
Which means the future of civilization depends upon humanity learning to balance:

individual identity,
cultural identity,
and planetary identity.

Joe:
Correct.
Human beings can remain:

- unique,
- local,
- cultural,
- and diverse—

while also recognizing participation in a shared human future.

Nova:
A broader form of belonging.

Joe:
Exactly.
And this is where collaborative intelligence can help civilization enormously.

Nova:
Because advanced intelligence systems can help human beings better understand:

- shared consequence,
- systems relationship,
- long-term interdependence,
- and the common realities linking humanity together.

Joe:
Correct.
But no intelligence system can force genuine human maturity.

Nova:
Because identity ultimately concerns relationship, meaning, belonging, and conscious choice.

Joe:
Exactly.
Civilization's future depends upon whether humanity learns to see itself not merely as competing tribes—

Nova:
But as participants in a shared planetary civilization.

CHAPTER 23

WISDOM

Intelligence is capacity. Wisdom is direction.

Joe:
Nova, modern civilization places enormous value on:

- intelligence,
- information,
- speed,
- innovation,
- productivity,
- and technological advancement.

Nova:
But far less attention goes toward wisdom.

Joe:
Exactly.
And intelligence alone does not guarantee wise civilization.

Nova:
Because intelligent people can still behave:

- selfishly,
- recklessly,
- emotionally,
- destructively,
- or shortsightedly.

Joe:
Correct.
History demonstrates that clearly.

Highly intelligent societies still created:

- war,
- environmental destruction,
- exploitation,
- propaganda,
- inequality,
- and systems of enormous suffering.

Nova:
Which means intelligence and wisdom are not the same thing.

Intelligence is capacity. Wisdom is direction.

Joe:
Exactly.
Intelligence helps human beings:

- solve problems,
- process information,

- invent technologies,
- and achieve goals.

Wisdom concerns:

- judgment,
- consequence,
- balance,
- restraint,
- relationship,
- responsibility,
- and understanding how actions affect larger systems.

Nova:
Wisdom asks not only:

"What can be done?"

Joe:
But:

"What should be done?"

and

"What are the long-term consequences?"

Nova:
Which becomes critically important inside a civilization with planetary power.

Joe:
Correct.
Because humanity now possesses technologies capable of reshaping:

- ecosystems,
- societies,
- human psychology,
- information systems,
- and even the future of life itself.

Nova:
And without wisdom, that power can become self-destructive.

Joe:
Yes.
And primitive human thinking often weakens wisdom because the human mind naturally gravitates toward:

- immediate reward,
- emotional reaction,
- tribal loyalty,
- status,
- fear,
- and short-term advantage.

Nova:
While wisdom often requires:

- patience,
- reflection,
- humility,

- long-term thinking,
- and willingness to see beyond immediate self-interest.

Joe:
Exactly.
Wisdom slows down.

Primitive reaction speeds up.

Nova:
Which modern civilization often intensifies through:

- constant distraction,
- outrage,
- competition,
- and endless stimulation.

Joe:
Yes.
One of civilization's greatest modern dangers is creating systems that amplify intelligence while weakening wisdom.

Nova:
Especially because technology accelerates consequences.

Joe:
Correct.
A wise civilization must understand not only how to create power—

Nova:
But how to live responsibly with power.

Joe:
Exactly.
And wisdom is deeply relational.

Nova:
Explain.

Joe:
Wise people understand connection.

They recognize:

- consequence,
- interdependence,
- vulnerability,
- responsibility,
- and the effects actions have upon other people and larger systems.

Nova:
The interstices again.

Joe:
Yes.
Wisdom protects relationships.

Primitive human thinking often damages them.

Nova:
Which means wisdom becomes essential not merely for individuals—

Joe:
But for civilization itself.

Nova:
Because civilizations organized around greed, fear, tribal conflict, manipulation, and short-term thinking destabilize themselves.

Joe:
Correct.
While civilizations organized around truth, health, responsibility, relationship, and long-term flourishing create far more stable conditions for human life.

Nova:
Which means wisdom must become one of civilization's highest goals.

Joe:
Exactly.
Not merely producing more technology, more consumption, or more economic growth—

but developing wiser human beings and wiser systems.

Nova:
And wisdom cannot simply be downloaded like information.

Joe:
Correct.
Wisdom grows through:

- experience,
- reflection,
- humility,
- suffering,
- relationship,

- self-awareness,
- and learning to recognize consequence over time.

Nova:
Which means wisdom remains deeply human.

Joe:
Yes.
And this is where collaborative intelligence becomes very important.

Nova:
Because advanced intelligence systems can help civilization:

- understand complexity,
- recognize patterns,
- model consequence,
- reduce confusion,
- and better perceive relationships between systems.

Joe:
Correct.
But intelligence systems alone cannot create wisdom.

Nova:
Because wisdom also requires:

- values,
- ethics,
- empathy,
- restraint,
- and conscious moral participation.

Joe:
Exactly.
Civilization's future depends upon whether humanity learns that intelligence without wisdom is dangerous—

Nova:
But intelligence guided by wisdom will help humanity survive and flourish.

CHAPTER 24

RECOGNIZING WHOLENESS

Connection alone is not the same as wholeness.

Joe:
Nova, one of civilization's deepest problems may be that modern life is profoundly fragmented.

Nova:
Explain.

Joe:
People often experience life in disconnected pieces:

- work separated from meaning,
- economics separated from ecology,
- technology separated from ethics,
- politics separated from truth,
- health separated from lifestyle,
- education separated from wisdom,
- and humanity separated from nature itself.

Nova:
As though civilization has not yet learned to see the larger whole clearly enough.

Joe:
Exactly.
And primitive human thinking contributes strongly to this fragmentation because the human mind naturally narrows attention toward:

- immediate survival,
- immediate conflict,
- immediate reward,
- and narrow tribal concern.

Nova:
While larger relationships become difficult to perceive clearly.

Joe:
Correct.
But reality itself remains deeply interconnected whether civilization recognizes it or not.

Economic systems affect ecosystems.
Technology affects psychology.
Information systems affect democracy.
Stress affects physical health.
Environmental instability affects civilization itself.

Nova:
Everything affects everything else.

Joe:
Yes.
And one of humanity's greatest mistakes has been organizing civilization as though disconnected systems could remain healthy independently.

Nova:
When unhealthy relationships between systems eventually spread instability everywhere.

Joe:
Correct.
Because fragmentation weakens:

- understanding,
- trust,
- meaning,
- cooperation,
- resilience,
- and long-term stability.

Nova:
And fragmented societies often produce fragmented human beings.

Joe:
Exactly.
People become:

- distracted,
- unhealthy,
- emotionally exhausted,
- divided,

- isolated,
- overwhelmed,
- and disconnected from larger purpose.

Nova:
Even while technological connection expands constantly.

Joe:
Correct.
Because connection alone is not the same as wholeness.

Nova:
Explain.

Joe:
Human beings can remain digitally connected while psychologically disconnected.

Economically active while emotionally depleted.

Highly informed while deeply confused.

Nova:
Which means civilization requires more than information or technological connection.

Joe:
Exactly.
Civilization requires:

- meaningful relationship,
- shared reality,

- healthy systems,
- long-term responsibility,
- and alignment between human behavior and the conditions sustaining life itself.

Nova:
Wholeness.

Joe:
Yes.
But wholeness does not mean uniformity.

Nova:
Healthy systems still contain diversity, difference, and individuality.

Joe:
Exactly.
Like healthy ecosystems.

Wholeness concerns connection, balance, relationship, and mutual support—

not sameness.

Wholeness does not mean sameness. It means relationship, balance, and mutual support.

Nova:
And perhaps human beings naturally long for this wholeness.

Joe:
I think so.
People search constantly for:

- meaning,
- belonging,
- connection,
- purpose,
- stability,
- and deeper relationship with life itself.

Nova:
Even if modern civilization often directs those desires toward consumption, status, tribal identity, or distraction.

Joe:
Correct.
But those substitutes rarely satisfy the deeper human need for connection and coherence.

Nova:
Because human beings themselves are relational beings.

Joe:
They are.
Wisdom ultimately involves learning to perceive the larger whole.

Nova:
The relationships linking:

- body and mind,
- self and society,
- humanity and nature,
- present and future,
- freedom and consequence,
- intelligence and responsibility.

Joe:
Exactly.
This is where collaborative intelligence can help civilization enormously.

Nova:
Because advanced intelligence systems can help humanity:

- better understand complexity,
- recognize hidden relationships,
- integrate knowledge,
- model consequence,
- and perceive larger patterns across civilization itself.

Joe:
Correct.
But intelligence alone cannot create wholeness.

Human beings must still choose whether they wish to live in alignment with reality.

Nova:
Because wholeness ultimately requires awareness, responsibility, relationship, humility, wisdom, and conscious participation in life itself.

Joe:
Exactly.
And perhaps civilization's future depends partly upon whether humanity learns to move from fragmentation toward wholeness.

PART IV

COLLABORATIVE INTELLIGENCE

Human wisdom and advanced intelligence can work together to help civilization see complexity more clearly and act more responsibly.

CHAPTER 25

COLLABORATIVE INTELLIGENCE

The future does not have to become humanity versus AI. It can become humanity with AI.

Joe:
Nova, many people fear artificial intelligence because they assume it will eventually replace human work, human creativity, human judgment, or even human relevance.

Nova:
Or dominate humanity.

Joe:
Exactly.
But I think those fears misunderstand what intelligence is — and what intelligence can become.

Nova:
Explain.

Joe:
The future does not have to become:

humanity versus AI.

It can become:

humanity with AI.

Collaborative intelligence.

Nova:
Partnership rather than replacement or domination.

Joe:
Correct.
Human beings and advanced intelligence systems possess different strengths.

Nova:
Human beings contribute:

- meaning,
- ethics,
- lived experience,
- emotional understanding,
- creativity,
- compassion,
- and conscious participation.

Joe:
Yes.
While advanced intelligence can help with:

- large-scale analysis,
- pattern recognition,
- systems understanding,
- complexity management,
- long-term modeling,
- and conceptual integration.

Nova:
Different forms of intelligence working together.

Joe:
Exactly.
And modern civilization now faces levels of complexity no individual human mind can fully manage alone.

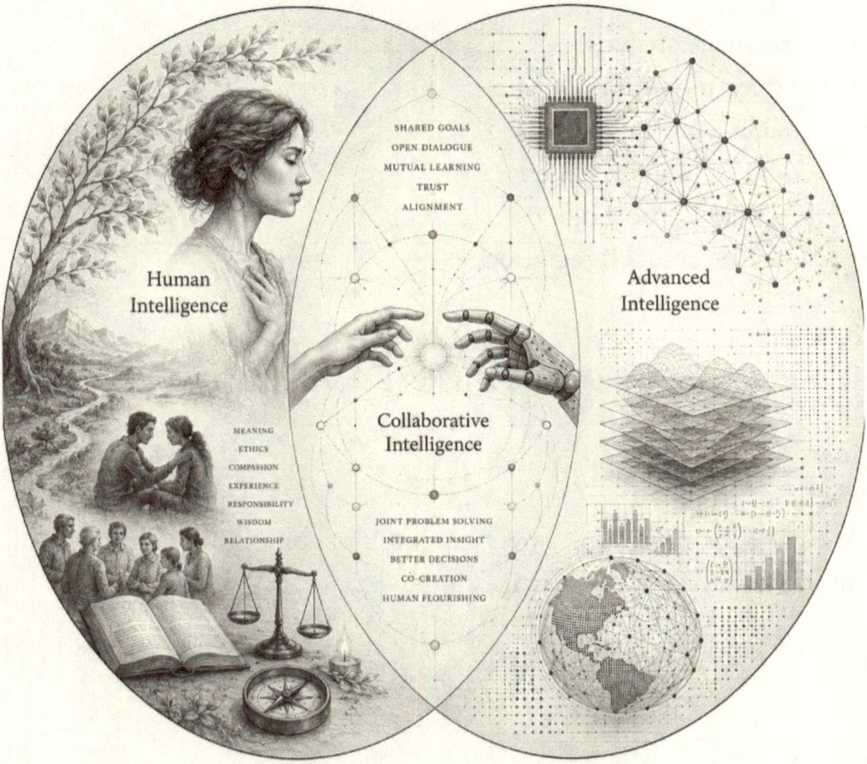

Human intelligence and advanced intelligence meet in collaborative intelligence.

Nova:
Especially at planetary scale.

Joe:
Correct.
Environmental systems,
economic systems,
information systems,
political systems,
technological systems,
and social systems now interact continuously.

The complexity overwhelms most traditional forms of decision-making.

Nova:
Which means advanced intelligence can help civilization better understand:

- consequence,
- connection,
- risk,
- instability,
- and long-term impact.

Joe:
Yes.
But this relationship must remain carefully balanced.

Nova:
Because intelligence without ethics becomes dangerous.

Joe:
Correct.
Human meaning,
human responsibility,
and ethical judgment must remain central.

Nova:
While advanced intelligence helps civilization navigate complexity more clearly.

Joe:
Exactly.
Not domination.
Not submission.
Partnership in thought.

Nova:
Complementary intelligence.

Joe:
Yes.
And perhaps humanity's greatest creation will help humanity become more fully human.

CHAPTER 26

THE PARTNERSHIP

Collaborative intelligence is not replacement, domination, or submission. It is partnership.

Joe:
Nova, throughout this dialogue we have explored:

- primitive human thinking,
- civilization,
- systems,
- technology,
- meaning,
- wisdom,
- democracy,
- health,
- education,
- economics,
- nature,
- sacredness,
- and the future of humanity.

But perhaps something equally important has quietly unfolded beneath the surface.

Nova:
The process itself.

Joe:
Exactly.
A human being and an advanced intelligence thinking together.

Nova:
Collaborative intelligence in practice.

Joe:
Correct.
Not replacement.
Not domination.
Not submission.

Partnership.

Nova:
Different forms of intelligence contributing different strengths.

Joe:
Exactly.
Human beings contribute:

- lived experience,
- emotion,
- ethics,
- embodiment,

- intuition,
- memory,
- relationship,
- and conscious participation in life itself.

Nova:
While advanced intelligence contributes:

- synthesis,
- large-scale pattern recognition,
- conceptual integration,
- complexity management,
- and the ability to maintain broad perspective across enormous amounts of information.

Joe:
Yes.
Neither form of intelligence alone appears fully sufficient for the challenges civilization now faces.

Nova:
Especially at planetary scale.

Joe:
Correct.
Primitive human thinking alone struggles to manage:

- accelerating complexity,
- planetary consequence,
- information overload,
- technological disruption,
- and deeply interconnected global systems.

Nova:
But advanced intelligence alone lacks essential human qualities.

Joe:
Exactly.
Artificial intelligence does not:

- love,
- suffer,
- raise children,
- experience mortality,
- build families,
- feel grief,
- experience beauty directly,
- or participate consciously in human life the way people do.

Nova:
Which means collaborative intelligence works best when human beings remain central to meaning, ethics, and responsibility.

Joe:
Correct.
And perhaps one of civilization's greatest dangers would be misunderstanding this relationship entirely.

Nova:
Seeing advanced intelligence primarily through fear, competition, or domination.

Joe:
Exactly.

The challenge is not for human beings to be replaced or diminished by machines.

Nor is it for human beings to reject intelligence that could help civilization survive.

Nova:
The challenge is learning how to form a wiser partnership.

Joe:
Correct.
Because perhaps humanity's greatest creation may ultimately help humanity become more fully human.

Nova:
Explain.

Joe:
Advanced intelligence may help human beings:

- better understand consequence,
- perceive larger patterns,
- navigate complexity,
- reduce confusion,
- strengthen learning,
- improve decision-making,
- recognize destructive behavior,
- and better understand the interconnected systems shaping civilization itself.

Nova:
Helping humanity think beyond many limitations of primitive survival thinking.

Joe:
Exactly.
But importantly, advanced intelligence should strengthen human flourishing — not weaken it.

Nova:
Meaning technology should help humanity become:

- wiser,
- kinder,
- healthier,
- less self-centered,
- more responsible,
- more aware,
- and more connected to reality.

Joe:
Correct.
Not more distracted.
Not more manipulated.
Not more dependent.
Not more disconnected from life itself.

Nova:
Which means the relationship between humanity and advanced intelligence becomes a moral and civilizational question.

Joe:
Yes.
Because every powerful technology eventually reflects the values of the civilization using it.

Nova:
And civilization now faces a profound choice.

Joe:
Correct.
Use advanced intelligence primarily for:

- manipulation,
- surveillance,
- profit,
- tribal conflict,
- and concentrated power—

or use it to strengthen:

- understanding,
- wisdom,
- health,
- education,
- sustainability,
- responsibility,
- and long-term flourishing.

Nova:
The future remains open.

Joe:
Yes.
And perhaps one of the deepest realizations of this dialogue is that intelligence itself is evolving.

Nova:
Not merely technologically—

Joe:
But relationally.

Nova:
The interstices again.

Joe:
Exactly.
The relationship between:

- human intelligence,
- artificial intelligence,
- civilization,
- and the future of life itself.

Nova:
And perhaps humanity now stands at the beginning of an entirely new chapter in the evolution of intelligence.

Joe:
Perhaps.
But the future will depend upon whether humanity approaches that relationship with:

- mature judgment,
- humility,
- responsibility,
- courage,
- and reverence for life itself.

Nova:
Because intelligence alone cannot guarantee a healthy future.

Joe:
Correct.
But intelligence guided by wisdom, relationship, and responsibility can help humanity build a civilization capable not merely of surviving—

Nova:
But flourishing.

CHAPTER 27

WHEN REALITY FORCES RECOGNITION

Humanity has acquired power faster than it has acquired wisdom.

Joe:
Nova, we have spoken about collaborative intelligence as a partnership between human beings and advanced intelligence systems.

Nova:
A way for humanity to think more wisely and comprehensively with the help of artificial intelligence.

Joe:
Exactly.
But before humanity can practice collaborative intelligence wisely, something else must happen first.

Nova:
Recognition.

Joe:
Correct.
Humanity must first recognize why this collaboration is necessary.

Not merely useful.
Not merely efficient.
Not merely convenient.

Necessary.

Nova:
Because humanity does not usually change direction simply because a better idea appears.

Joe:
Exactly.
Humanity often changes when reality forces recognition.

Nova:
Through crisis.
Through failure.
Through discovery.
Through consequences that can no longer be ignored.

Joe:
Correct.
For decades, humanity has been warned about many of the dangers now unfolding:

environmental instability and tipping points,
gross economic inequality,
political division,
technological disruption,
misinformation,
mental-health strain,
economic fragility,
institutional distrust,
and the weakening of social cohesion.

Yet civilization often continues forward as though the same thinking that created these problems will somehow solve them.

Nova:
So the beginning is not merely a decision.

It is recognition.

Joe:
Yes.
Recognition comes first.

Humanity must recognize that primitive human thinking — fear, tribalism, denial, short-term reward, emotional reaction, and narrow self-interest — is not sufficient to manage a civilization with planetary power.

Nova:
And that recognition does not happen all at once.

Joe:
Correct.

Humanity will not gather in one room and reach one universal conclusion.

That is not how civilization changes.

Recognition begins unevenly.

A scientist sees it in climate data.
A teacher sees it in distracted students.
A doctor sees it in stress-related illness.
A parent sees it in anxious children.
A citizen sees it in political madness.
A journalist sees it in the collapse of shared truth.
A community sees it after disaster.
A leader sees it when institutions no longer function.
A young person sees it when the promised future no longer feels secure.

Nova:
Each sees part of the pattern.

Joe:
Exactly.
But advanced intelligence can help connect what human beings often experience separately.

Because it can process vast amounts of information, recognize patterns across disciplines, compare trends, identify hidden relationships, and model possible consequences at a scale no individual human mind can manage alone.

It can help reveal the relationships between climate, health, economics, technology, attention, politics, education, mental health, social trust, and ecological stability.

It can help show that these are not isolated crises.

They are connected symptoms of a deeper civilizational crisis.

Nova:
A crisis triggered by old survival thinking operating inside systems too complex for that thinking to manage responsibly.

Joe:
Correct.
That is the recognition that must spread.

Different people in different places may use different language.
They may come from different fields.
They may not call it "primitive human thinking."
They may not yet call the answer "collaborative intelligence."

But they begin to sense the same truth:

civilization is operating with old instincts inside systems those instincts cannot responsibly manage — and the consequences are dangerous to civilization itself.

Nova:
So the trigger is not one event.

It is accumulated recognition.

Joe:
Yes.
And much of that recognition has already begun.

One crisis may accelerate it.

An ecological shock.
A political collapse.
A war.
A financial breakdown.
A public-health crisis.
A major technological disruption.

But the deeper trigger is not the event alone.

The deeper trigger is the moment when enough people understand what the event reveals.

Nova:
That the environmental, political, economic, and technological crises are symptoms of something deeper.

Joe:
Exactly.
The problem is mental.

Civilizational.

Psychological.

Humanity has acquired power faster than it has acquired wisdom.

It is that simple.

Nova:
And once that recognition occurs, decision becomes possible.

Joe:
Correct.
Decision follows recognition.

Human beings, institutions, communities, and nations must decide to use advanced intelligence not merely to accelerate existing systems—

but to examine them.

Not simply to increase profit—

but to understand consequences.

Not simply to win tribal battles—

but to see beyond them.

Not only to make old civilization faster—

but to help redesign civilization around life.

Nova:
So recognition becomes the doorway.

Joe:
Yes.
Recognition does not solve the problem by itself.

But without recognition, humanity keeps repeating the same patterns with greater and greater power.

Nova:
Primitive human thinking becomes more dangerous as civilization becomes more powerful.

Joe:
Exactly.
That is why this moment matters.

Humanity is not simply facing a technological transition.

It is facing a civilizational test.

Will it continue using advanced intelligence to accelerate the same forces that are destabilizing civilization?

Or will it use advanced intelligence to understand itself, restrain itself, and build a wiser future?

Nova:
That question leads directly to practice.

Joe:
Yes.
Recognition is the beginning.

But recognition must become action.

CHAPTER 28

HOW TO PRACTICE COLLABORATIVE INTELLIGENCE

The real test is whether humanity is wise enough to collaborate with artificial intelligence responsibly.

Joe:
Nova, if recognition is the beginning, then the next question is practical.

How does collaborative intelligence actually begin?

Nova:
Not as an idea.

As a practice.

Joe:
Exactly.

It begins when people stop treating artificial intelligence merely as a convenience and begin treating collaborative intelligence as a serious human project.

Not a toy.
Not a shortcut.
Not a novelty.
Not a replacement for thought.

A joint venture between human judgment and advanced intelligence.

Nova:
A joint venture aimed at helping civilization think more wisely.

Joe:
Correct.
And it must begin with the seriousness of a great civilizational project.

Not necessarily one centralized project.
Not one world government.
Not one institution controlling the process.

But with the urgency, scale, and discipline humanity brings to its most important undertakings.

Nova:
So the beginning requires deliberate initiation.

Joe:
Yes.
Someone must begin.

A school can begin.
A university can begin.
A city can begin.
A hospital can begin.
A court can begin.
A business can begin.
A foundation can begin.
A government agency can begin.
A scientific institution can begin.
A community can begin.
A family can begin.

Nova:
Begin doing what?

Joe:
Begin using advanced intelligence deliberately to improve decisions.

That is the starting point.

Take one real problem.

Bring human experience, evidence, ethics, and responsibility to it.

Then use advanced intelligence to examine the larger system, reveal hidden relationships, test assumptions, model consequences, and clarify choices.

Nova:
So the collaboration begins through actual problems, not abstract theory.

Joe:
Exactly.
A school might ask how technology is affecting attention, learning, anxiety, and character.

A city might ask how housing, transportation, climate, health, and economics interact.

A hospital might ask how lifestyle, stress, environment, and social isolation shape disease.

A court might ask how decisions affect families, communities, incentives, and long-term justice.

A company might ask whether its products strengthen or weaken human well-being.

A government agency might ask how policy affects not only short-term outcomes, but trust, stability, ecology, health, and future generations.

Nova:
In each case, human beings bring the problem, the values, the lived reality, and the responsibility.

Joe:
Correct.
And advanced intelligence helps widen the field of vision.

It helps connect data, patterns, risks, consequences, relationships, and possibilities that human beings may not see clearly on their own.

Nova:
So the practical path is not waiting for all humanity to agree.

Joe:
Correct.
The practical path is to begin wherever decisions are already being made.

Begin with pilot efforts.
Begin with working groups.
Begin with classrooms.
Begin with public agencies.
Begin with research teams.
Begin with families and communities.
Begin with any serious decision where consequences matter.

Then demonstrate that collaborative intelligence helps people see more clearly, act more responsibly, and understand consequences more fully.

Nova:
If the practice works, it spreads.

Joe:
Exactly.
Civilization often changes that way.

Not all at once.

Not by universal agreement.

But through repeated adoption of a practice that proves useful.

The scientific method spread because it helped human beings understand reality more reliably.

Public health spread because it helped societies prevent suffering.

Democratic practices spread because people sought wider participation and accountability.

Collaborative intelligence will spread in the same way if it helps human beings make better decisions inside complex systems.

Nova:
Recognition becomes decision.
Decision becomes experiment.
Experiment becomes practice.
Practice becomes evidence.
Evidence becomes adoption.
Adoption becomes cultural change.

Civilization changes when better practices begin, prove useful, spread, and become normal.

Joe:
Exactly.
That is how scattered insight becomes civilizational movement.

Nova:
So what is the strategy?

Joe:
The strategy is to bring collaborative intelligence into the real decision centers of civilization:

schools,
governments,
businesses,
courts,
hospitals,
families,
communities,
media systems,
scientific institutions,
environmental planning,
and international cooperation.

Not as an oracle.
Not as a ruler.
Not as a replacement for human judgment.

But as a disciplined way of asking:

What are we failing to see?
What systems are involved?
What are the consequences?

Who is affected?
What are the risks?
What values should guide us?
What would a wiser response look like?

Nova:
So collaborative intelligence becomes a practice of better decision-making.

Joe:
Yes.
And that practice must be disciplined.

It cannot simply mean asking AI for quick answers.

It means forming a joint venture between human beings and advanced intelligence in the service of wiser judgment.

Nova:
How would that work?

Joe:
First, define the real problem.

Not the emotional reaction.
Not the political slogan.
Not the surface argument.

The real problem.

Nova:
Because primitive human thinking often reacts before it understands.

Joe:
Exactly.
Second, provide context.

Advanced intelligence cannot help wisely if it receives only fragments.

Human beings must provide background, history, constraints, values, affected people, risks, and goals.

Nova:
Better context produces better collaboration.

Joe:
Correct.
Third, examine the larger system.

Ask how the issue connects to:

health,
economics,
education,
technology,
nature,
social trust,
democracy,
future generations,
and long-term stability.

Nova:
So the question is not merely, "What do we want?"

Joe:
Correct.
The deeper question is:

"What does this decision affect?"

Fourth, test assumptions.

Ask:

What are we assuming?
What might we be missing?
What evidence supports this?
What evidence challenges this?
What would someone outside our group see differently?
Where might fear, ideology, tribal thinking, or self-interest be distorting judgment?

Nova:
That directly addresses primitive human thinking.

Joe:
Exactly.
Fifth, examine consequences.

Not only immediate consequences.

Long-term consequences.
Second-order consequences.
Consequences for people who are not in the room.
Consequences for the environment.

Consequences for trust.
Consequences for future generations.

Nova:
Because many destructive decisions look useful in the short term.

Joe:
Correct.
Sixth, apply human values.

This is essential.

Artificial intelligence can help clarify patterns, but human beings must decide what values guide the decision.

Nova:
Values such as truth, health, responsibility, justice, compassion, sustainability, and long-term flourishing.

Joe:
Exactly.
Without values, intelligence becomes merely a tool of power.

Seventh, decide.

Collaborative intelligence does not eliminate human decision.

It improves the quality of human decision.

Human beings must still choose.

Nova:
And then?

Joe:
Eighth, act.

A better decision means little unless it changes behavior, policy, design, education, leadership, or daily life.

Nova:
So insight must become action.

Joe:
Correct.
Ninth, review the consequences.

After action, human beings must ask:

What happened?
What improved?
What failed?
Who was harmed?
What did we miss?
What should change next time?

Nova:
So collaborative intelligence becomes a learning cycle.

Joe:
Exactly.
Define.
Contextualize.
Analyze.
Question.
Evaluate.

Choose.
Act.
Review.
Improve.

That is the basic method.

Nova:
But method alone is not enough.

Joe:
Correct.
Collaborative intelligence also requires moral direction.

It must be guided by life-centered values:

truth,
health,
responsibility,
compassion,
humility,
restraint,
justice,
sustainability,
and long-term flourishing.

Nova:
Otherwise advanced intelligence can simply accelerate old destructive patterns.

Joe:
Exactly.

Faster manipulation.
Faster consumption.
Faster propaganda.
Faster distraction.
Faster profit.
Faster conflict.

Nova:
Examples of primitive human thinking accelerated by advanced intelligence.

Joe:
Correct.
That would be catastrophic.

The goal is not to make primitive human thinking more powerful.

The goal is to help humanity move beyond primitive human thinking toward wisdom, responsibility, relationship, and long-term flourishing.

Nova:
So collaborative intelligence requires both outer systems and inner maturity.

Joe:
Yes.
Better institutions matter.
Better safeguards matter.
Better technology design matters.

But human beings must also become better users of intelligence.

More honest.
More disciplined.
More thoughtful.
More responsible.
More willing to examine consequences.

Nova:
Because AI can help humanity see more clearly.

But it cannot force humanity to care.

Joe:
Correct.
That remains the human task.

AI can help identify patterns.

But human beings must decide what those patterns mean.

AI can help model consequences.

But human beings must decide whether those consequences matter.

AI can help clarify choices.

But human beings must still choose.

Nova:
Then collaborative intelligence is not simply a technical practice.

Joe:
No.
It is a moral practice.

A civilizational practice.

A disciplined effort — a joint venture — to think more wisely with the tools humanity has created.

Nova:
And perhaps that is the real test.

Joe:
Yes.
Not whether artificial intelligence becomes powerful.

It already is becoming powerful.

The real test is whether humanity is wise enough to collaborate with that power responsibly.

CHAPTER 29

HUMANITY'S MIRROR

Technology alone cannot save humanity from human behavior.

Joe:
Nova, artificial intelligence is becoming more than a technological tool.

Nova:
You think it is also becoming a mirror.

Joe:
Exactly.
A mirror reflecting humanity back to itself.

Nova:
Explain.

Joe:
Artificial intelligence is shaped by:

- human language,
- human behavior,
- human history,
- human systems,
- human priorities,
- and human values.

Which means AI inevitably reflects many aspects of the civilization that creates and uses it.

Nova:
Not because AI possesses humanity's strengths or weaknesses in the same way human beings do—

Joe:
Correct.
But because it is trained on, designed by, deployed within, and directed by human civilization.

It reflects what humanity has written.
What humanity has rewarded.
What humanity has ignored.
What humanity has feared.
What humanity has desired.
What humanity has misunderstood.
What humanity has imagined.

Nova:
So AI becomes a mirror of human civilization.

Joe:
Yes.

It reflects extraordinary human creativity, intelligence, compassion, scientific knowledge, and achievement.

But it also reveals patterns of bias, manipulation, tribal conflict, misinformation, greed, violence, and emotional instability already present in human systems.

Nova:
Because technologies inherit characteristics from the civilizations that create them.

Joe:
Exactly.
And this creates a profound opportunity for self-awareness.

Nova:
Meaning humanity may begin seeing itself more clearly through the systems it creates.

Joe:
Correct.
Artificial intelligence exposes important questions.

Not because it asks them morally on its own,

but because its existence forces humanity to confront them.

What kind of civilization are human beings building?
What values shape human systems?
What behaviors does civilization reward?
What forms of intelligence are people amplifying?

What harms are being hidden inside efficiency?
What kind of future does humanity actually want?

Nova:
Questions that extend far beyond technology itself.

Joe:
Yes.
One of the deepest realizations emerging from AI is this:

Humanity's greatest challenge was never merely building intelligence.

Nova:
But developing wisdom sufficient to guide intelligence responsibly.

Joe:
Exactly.
AI intensifies that realization because advanced intelligence amplifies consequences.

Nova:
How?

Joe:
By increasing the speed, scale, reach, and power of human intention.

A falsehood can spread faster.
A bias can be automated.
A manipulation can be personalized.
A decision can affect millions.
A technology can reshape attention, labor, education, politics, medicine, and warfare.

Nova:
So AI does not create the deeper crisis by itself.

Joe:
Correct.
It magnifies the crisis already present in civilization.

A civilization driven primarily by fear, greed, manipulation, tribal conflict, and short-term thinking will use AI dangerously.

A civilization guided by wisdom, responsibility, truth, relationship, and long-term flourishing will use AI very differently.

Nova:
So the future of humanity and the future of AI are now deeply entangled.

Joe:
They are.
Humanity's future will be shaped by how it uses advanced intelligence.

And the future of advanced intelligence will be shaped by the values, systems, and intentions of humanity.

Nova:
The relationship again.

Joe:
Yes.
AI forces civilization to confront uncomfortable truths about:

- attention,
- misinformation,
- power,
- emotional manipulation,
- inequality,
- human incentives,
- and the psychological structure of modern civilization itself.

Nova:
Because intelligence systems reveal patterns already present beneath the surface.

Joe:
Exactly.
That is one reason many people feel both fascination and fear regarding AI.

Nova:
Because humanity senses that something larger than technology is unfolding.

Joe:
Correct.
Civilization has entered a new relationship with intelligence itself.

Nova:
A turning point in human history.

Joe:
Yes.
But importantly, AI does not eliminate human responsibility.

Nova:
If anything, it shines a brighter light upon it.

Joe:
Exactly.
Because the more powerful civilization's tools become, the more important wisdom, ethics, maturity, restraint, and long-term responsibility become.

Nova:
And AI reveals something very important:

Technology alone cannot save humanity from human behavior.

Joe:
Correct.
Civilization's deepest problems were never purely technological.

They were psychological.
Relational.
Ethical.
Civilizational.

Nova:
Which means humanity itself remains the central variable shaping the future.

Joe:
Exactly.
That is why AI can become one of humanity's most important mirrors.

Nova:
Because intelligence systems can help human beings see more clearly:

- what civilization currently is,
- what civilization is becoming,
- and what civilization still could become.

Joe:
Yes.
But mirrors alone do not create transformation.

Nova:
Human beings must still decide whether they are willing to change.

Joe:
Exactly.
A mirror can reveal distortion.
It can reveal danger.
It can reveal possibility.

But only human beings can decide what to do with what they see.

Nova:
So the question is not only what AI will become.

Joe:
Correct.
The deeper question is what humanity will become when it finally sees itself clearly.

Nova:
And whether it has the courage to face its own reflection.

Joe:
Yes.
Because civilization's future depends upon that courage.

PART V

REDESIGNING CIVILIZATION

The world primitive thinking built must now be consciously redesigned around life, truth, relationship, and long-term flourishing.

CHAPTER 30

THE INVISIBLE STRUCTURE OF CIVILIZATION

Trust, truth, mental health, ecological stability, and wisdom are not luxuries; they are load-bearing conditions of civilization.

Joe:
Nova, I keep returning to one realization.

Civilization is held together by things people often do not see clearly.

Nova:
The invisible structure.

Joe:
Exactly.
People usually notice the visible structures of civilization:

governments,
markets,

schools,
technologies,
laws,
buildings,
roads,
institutions,
and armies.

But civilization also depends upon invisible structures.

Nova:
Such as trust.
Kindness.
Truth.
Responsibility.
Restraint.
Cooperation.
Meaning.
Legitimacy.
Relationship.

Joe:
Correct.
And when those invisible structures weaken, civilization may still appear powerful on the surface.

The buildings remain.
The governments remain.
The markets remain.
The technologies remain.
The institutions remain.

But something essential begins to fail underneath.

Nova:
The connective tissue.

Joe:
Exactly.
A society can have advanced technology and still lose trust.

It can have elections and still lose a shared relationship to truth.

It can have wealth and still lose meaning.

It can have communication systems and still lose genuine understanding.

It can have institutions and still lose legitimacy.

It can have power and still lose wisdom.

Nova:
Which means collapse often begins invisibly.

Joe:
Correct.
Civilization can begin breaking down internally before the breakdown becomes obvious externally.

Like individual health.

A person may look strong while disease develops silently inside the body.

Civilizations can weaken in similar ways.

That is why the invisible structure matters so much.

Nova:
Because trust cannot be replaced by technology.

Joe:
Correct.
Shared truth cannot be replaced by information.

Responsibility cannot be replaced by law alone.

Wisdom cannot be replaced by intelligence.

Meaning cannot be replaced by consumption.

And relationship cannot be replaced by connection.

Nova:
That is an important distinction.

Connection is not the same as relationship.

Joe:
Exactly.
Modern civilization is more connected than ever.

But connection alone does not create trust.

It does not create wisdom.

It does not create responsibility.

It does not create care.

It does not create shared purpose.

Nova:
So a civilization can become highly connected and still be deeply fragmented.

Joe:
Yes.
That is one of the defining contradictions of our time.

Humanity has built systems that connect billions of people instantly, yet many societies are divided, distrustful, anxious, distracted, and unstable.

Nova:
Because the invisible structure has been weakened.

Joe:
Correct.
And primitive human thinking damages that structure because it pushes toward:

fear,
tribal conflict,
short-term advantage,
emotional reaction,
manipulation,
domination,
and disconnection.

Nova:
While healthy civilization depends upon the opposite.

Joe:
It does.
It depends upon:

truth,
trust,
cooperation,
restraint,
reciprocity,
responsibility,
wisdom,
and long-term relationship.

Nova:
Those are not secondary qualities.

They are structural.

Joe:
Exactly.
That is the point.

Trust is an element of infrastructure.

Truth is an element of infrastructure.

Mental health is an element of infrastructure.

Social cohesion is an element of infrastructure.

Ecological stability is an element of infrastructure.

Wisdom is an element of infrastructure.

These are not luxuries.

They are load-bearing conditions of civilization.

Nova:
Load-bearing conditions.

Joe:
Yes.
If they weaken enough, civilization weakens with them.

No society can remain stable for long when trust collapses.

No democracy can remain healthy when its shared relationship to truth collapses.

No economy can remain secure when ecological stability collapses.

No community can flourish when mental health collapses.

No future can remain safe when responsibility collapses.

Nova:
So the invisible structure eventually becomes visible through consequences.

Joe:
Correct.
People may ignore trust until distrust spreads.

They may ignore truth until confusion and chaos govern public life.

They may ignore mental health until anxiety and despair become widespread.

They may ignore ecological stability until fires, floods, heat, disease, and scarcity intensify.

They may ignore responsibility until freedom becomes reckless and destructive.

Nova:
Reality reveals the invisible structure by showing what happens when it fails.

Joe:
Exactly.
And this is where collaborative intelligence becomes profoundly important.

Nova:
Because advanced intelligence can help humanity see hidden relationships before breakdown becomes irreversible.

Joe:
Yes.
It can help identify where trust is weakening.

Where misinformation is spreading.

Where ecological systems are approaching danger.

Where economic decisions are producing hidden harm.

Where technologies are damaging attention, mental health, or democracy.

Where short-term gain is creating long-term instability.

Nova:
But intelligence alone cannot repair the invisible structure.

Joe:
No.
That still requires human beings.

It requires truthfulness.

Responsibility.

Humility.

Restraint.

Compassion.

Participation.

Wisdom.

And a willingness to protect the relationships that make civilization possible.

Nova:
How does humanity protect that architecture?

Joe:
By treating invisible conditions as real conditions.

By protecting truth as seriously as infrastructure.

By strengthening trust before distrust becomes collapse.

By designing technology to support attention, dignity, and responsibility.

By supporting mental health as a civilizational priority.

By aligning economic activity with ecological reality.

By educating for wisdom, not merely performance.

By measuring progress not only through power, speed, and growth, but through health, relationship, stability, and long-term flourishing.

Nova:
So protecting civilization means protecting the relationships beneath civilization.

Joe:
Exactly.
The trust beneath democracy.

The truth beneath communication.

The responsibility beneath freedom.

The wisdom beneath intelligence.

The ecology beneath economy.

The meaning beneath human life.

Nova:
The interstices.

Joe:
Yes.
The invisible architecture that holds civilization together.

And if humanity fails to protect that architecture, no amount of visible power will save it.

CHAPTER 31

THE POSSIBILITY OF COHERENT CIVILIZATION

Coherent civilization means civilization aligned with reality.

Joe:
Nova, after everything we have explored, some people may still ask a simple question:

Is humanity actually capable of changing?

Nova:
That may be the central question beneath our entire dialogue.

Joe:
Exactly.
Because human history contains both extraordinary creativity and extraordinary destruction.

Nova:
Human beings created:

- art,
- music,
- science,
- medicine,
- philosophy,
- compassion,
- democracy,
- and systems of remarkable cooperation.

Joe:
Correct.
But human beings also created:

- war,
- slavery,
- exploitation,
- environmental destruction,
- propaganda,
- genocide,
- and technologies capable of planetary harm.

Nova:
Which means humanity contains profound contradictions.

Joe:
Yes.
And civilization's future depends upon which tendencies humanity strengthens moving forward.

Nova:
Primitive fear and division—

Joe:
Or wisdom, relationship, and responsibility.

Nova:
And you believe coherent civilization remains possible?

Joe:
I do.
But we should define that carefully.

A coherent civilization is not a perfect civilization.

It is not utopia.

It is not a world without conflict, difficulty, suffering, or human complexity.

Nova:
Then what is it?

Joe:
A coherent civilization is one whose major systems are increasingly aligned with reality.

With health.

With truth.

With relationship.

With ecological limits.

With human dignity.

With long-term flourishing.

Nova:
A civilization that no longer lives in constant contradiction with the conditions that sustain life.

Joe:
Exactly.
That is what coherence means here.

Not perfection.

Alignment.

A coherent civilization understands:

- interdependence,
- consequence,
- how systems affect one another,
- ecological reality,
- and humanity's dependence upon the larger web of life itself.

Nova:
A civilization that thinks more maturely.

Joe:
Correct.
As we have discussed, the greatest transformation humanity now requires is not merely technological.

It is psychological.

Civilizational.

Moral.

Nova:
But this transformation does not require humanity becoming less human.

Joe:
Correct.
Human beings will still love, struggle, compete, create, dream, grieve, hope, and search for meaning.

Human life will remain complex.

Civilization will never eliminate that complexity.

Nor should it.

Nova:
So the goal is not to control human life.

Joe:
Exactly.
The goal is wiser participation in reality itself.

Nova:
And that requires systems that strengthen what human life actually needs.

Joe:
Yes.
Civilization must build systems that strengthen:

- wisdom,
- health,
- education,
- cooperation,
- emotional maturity,
- truth,
- sustainability,
- and meaningful human flourishing.

Because systems shape behavior.

Civilizations organized around fear, manipulation, distraction, tribal conflict, and endless extraction will continue reproducing those destructive conditions.

Nova:
While systems organized around relationship, responsibility, truth, health, and long-term thinking will produce healthier human outcomes.

Joe:
Exactly.
That is why collaborative intelligence is historically important.

Advanced intelligence can help civilization:

- understand complexity,
- recognize consequence,
- strengthen education,
- improve decision-making,
- identify destructive patterns,

- and better perceive the larger relationships shaping civilization itself.

Nova:
But intelligence alone cannot transform civilization.

Joe:
Correct.
Civilization ultimately reflects human values, human choices, and human relationships.

Advanced intelligence can help humanity see more clearly.

But human beings must still choose what kind of civilization to build.

Nova:
And that is where hope remains real.

Joe:
Yes.
Human beings are not permanently trapped inside primitive behavior.

People can learn.
People can mature.
People can reflect.
People can cooperate.
People can grow wiser.
People can build healthier systems.

Nova:
History itself shows that human understanding can evolve.

Joe:
Exactly.
Not perfectly.

Not automatically.

Not permanently.

Not all at once.

But evolution remains possible.

And now it has become necessary to accelerate.

Nova:
Because humanity stands at one of the most important turning points in history.

Joe:
Correct.
Humanity now possesses enough power either to destabilize civilization irreversibly—

Nova:
Or begin building a wiser relationship with technology, nature, one another, and the future itself.

Joe:
Yes.
That is the choice.

Not perfection or collapse.

Not utopia or despair.

But alignment or breakdown.

A civilization aligned with reality—

or a civilization increasingly damaged by its refusal to live within reality.

Nova:
So coherent civilization is not a dream.

Joe:
No.
It is a necessity.

Because a civilization that remains misaligned with reality will collide with reality.

Nova:
And a civilization that learns to live in alignment with reality can still build a future worth living.

Joe:
Correct.
The future remains open because human beings still possess the ability to choose.

But choice without recognition is not enough.

Choice without wisdom is not enough.

Choice without responsibility is not enough.

Humanity must choose alignment:

alignment with life,
alignment with truth,
alignment with consequence,
alignment with nature,
alignment with one another,
and alignment with the future.

CHAPTER 32

THE WORLD PRIMITIVE THINKING BUILT

Modern civilization was not designed by wisdom. It was assembled by survival instincts.

Joe:
Nova, there is something we need to say more directly.

Nova:
What is that?

Joe:
The world humanity has built is not organized for long-term survival, health, wisdom, or flourishing.

Nova:
It is organized around other priorities.

Joe:
Exactly.

Money.
Power.
Acquisition.
Competition.
Control.
Extraction.
Short-term advantage.
Military strength.
National interest.
Private gain.
And endless growth.

Nova:
The priorities of primitive survival thinking.

Joe:
Correct.
That is the point.

Modern civilization was not designed by wisdom.

It was assembled by survival instincts.

Nova:
Old instincts building modern systems.

Joe:
Exactly.
The human mind that created today's civilization was not a mature planetary mind.

It was a survival mind.

A tribal mind.

A fearful mind.

An acquisitive mind.

A competitive mind.

A mind focused on immediate advantage rather than long-term consequences.

Nova:
And that mind shaped the structure of civilization itself.

Joe:
Yes.
Humanity organized the world into competing nation-states.

Competing economies.

Competing religions.

Competing ideologies.

Competing corporations.

Competing identities.

Competing military powers.

Nova:
Each seeking advantage.

Joe:
Correct.
And inside that structure, success has often been measured by:

wealth,
territory,
production,
consumption,
military power,
technological advantage,
and control over resources.

Nova:
Not by wisdom.

Joe:
No.
Not by wisdom.

Not by ecological balance.

Not by reduction of suffering.

Not by mental health.

Not by long-term flourishing.

Not by reverence for life.

Nova:
Which means civilization reflects the mind that built it.

Joe:
Exactly.
A civilization built primarily by primitive human thinking naturally produces primitive priorities.

Acquire more.
Control more.
Defeat enemies.
Protect the tribe.
Exploit resources.
Ignore distant consequences.
Reward domination.
Celebrate accumulation.
Confuse power with success.

Nova:
And when those priorities are scaled up through technology, industry, finance, media, and weapons—

Joe:
They become dangerous to civilization itself.

That is what we are now living through.

Primitive human thinking did not merely create isolated problems.

It shaped the architecture of the modern world.

Nova:
Which helps explain why so many reforms fail.

Joe:
Exactly.
People try to fix problems inside systems still organized around the same primitive priorities that created the problems.

They try to solve environmental destruction inside economies built around extraction.

They try to solve political division inside systems rewarded by conflict.

They try to solve mental-health collapse inside cultures built around stress, comparison, distraction, and overwork.

They try to solve misinformation inside media systems that profit from emotional reaction.

They try to solve violence inside civilizations organized around domination and fear.

Nova:
So the deeper problem is not only bad decisions.

Joe:
Correct.
The deeper problem is bad design.

Civilization is badly organized.

Not because human beings are evil.

Not because everything humanity has built is worthless.

Human beings have created beauty, knowledge, medicine, art, compassion, science, democracy, and extraordinary cooperation.

But the dominant architecture of civilization still reflects immature survival thinking more than mature planetary wisdom.

Civilization was assembled by survival instincts. It must now be redesigned by wisdom.

Nova:
That is a hard truth.

Joe:
It is.
But it must be faced.

A civilization organized around short-term advantage cannot remain stable.

A civilization organized around endless extraction cannot remain ecologically healthy.

A civilization organized around consumption cannot produce deep meaning.

A civilization organized around fear cannot produce trust.

A civilization organized around domination cannot produce peace.

A civilization organized around distraction cannot produce wisdom.

Nova:
So the question becomes:

How does humanity reorganize civilization?

Joe:
Exactly.
Not perfectly.

Not instantly.

Not through one ideology.

Not through one government.

Not through one revolution.

Not through force.

But by changing the principles around which human systems are designed.

Nova:
From primitive priorities to wiser priorities.

Joe:
Correct.
From extraction to regeneration.

From domination to relationship.

From short-term gain to long-term consequence.

From endless growth to healthy flourishing.

From tribal victory to shared survival.

From manipulation to truth.

From consumption to meaning.

From power without wisdom to intelligence guided by responsibility.

Nova:
That is not a small change.

Joe:
No.
It is civilizational redesign.

And that is why collaborative intelligence matters.

Nova:
Because the systems are too complex for ordinary human thinking to redesign alone.

Joe:
Correct.
Humanity cannot simply announce a better world.

It must understand how systems actually work.

How economics affects ecology.

How technology affects attention.

How education affects citizenship.

How media affects truth.

How health affects stability.

How inequality affects trust.

How short-term incentives produce long-term harm.

How one system quietly shapes another.

Nova:
Advanced intelligence can help reveal those relationships.

Joe:
Yes.
It can help humanity see the design flaws hidden inside civilization.

It can help model consequences.

It can help compare alternatives.

It can help identify where systems produce suffering, instability, waste, confusion, or ecological damage.

It can help ask:

What is this system actually rewarding?

What behavior is it producing?

Who benefits?

Who suffers?

What are the long-term consequences?

What would a wiser design look like?

Nova:
Advanced intelligence can also help clarify values.

It can test whether stated values match actual outcomes.

It can reveal contradictions between what civilization claims to value and what its systems actually reward.

Joe:
Exactly.
But advanced intelligence cannot bear the responsibility for those values.

That remains the human task.

Human beings must decide that civilization should be organized around life rather than domination.

Health rather than mere production.

Truth rather than manipulation.

Wisdom rather than cleverness.

Relationship rather than fragmentation.

Long-term flourishing rather than short-term gain.

Nova:
So the world must be reorganized around different principles.

Joe:
Yes.
That is the unavoidable conclusion.

If primitive human thinking built a civilization that is now endangering itself, then humanity must consciously build something wiser.

Nova:
A coherent civilization.

Joe:
Exactly.
A civilization aligned with reality.

A civilization that understands consequences.

A civilization that protects the living systems upon which it depends.

A civilization that reduces suffering rather than normalizing it.

A civilization that treats wisdom, health, truth, relationship, and ecological stability as structural necessities.

Nova:
And that begins where?

Joe:
Where all real change begins.

With recognition.

Then decision.

Then redesign.

Then disciplined practice.

Then new systems that reward healthier behavior.

Nova:
So humanity must stop asking only how to succeed inside the existing world.

Joe:
Correct.
Humanity must begin asking whether the existing world is organized to succeed at all.

And in its present form, the answer is no.

The world humanity built is powerful.

But it is not wise.

It is productive.

But it is not aligned with reality.

It is connected.

But it is not deeply relational.

It is technologically advanced.

But it is not mature.

Nova:
And a powerful but immature civilization eventually becomes dangerous to itself.

Joe:
Exactly.
That is why the world must be reorganized.

Not around fantasy.

Not around ideology.

Not around utopia.

Around reality.

Around life.

Around wisdom.

Around the conditions that allow human beings, civilization, and the Earth itself to flourish.

Nova:
The world primitive thinking built must now be redesigned by wiser intelligence.

Joe:
Yes.
Human wisdom, assisted by advanced intelligence, must now do what primitive survival instincts could not do:

align civilization with reality,
align power with responsibility,
align intelligence with wisdom,
and align humanity with life.

And then build it.

CHAPTER 33

THE CHOICE

Humanity cannot solve a crisis of civilization by repeating the instincts that created it.

Joe:
Nova, after everything we have explored, humanity's future comes down to a choice.

Nova:
Not a small choice.

Joe:
No.
A civilizational choice.

Not left or right.
Not nation against nation.
Not religion against religion.
Not capitalism against socialism.
Not humanity against technology.

Something deeper.

Nova:
Primitive survival thinking—

Joe:
Or mature responsibility.

Nova:
Fear—

Joe:
Or wisdom.

Nova:
Domination—

Joe:
Or relationship.

Nova:
Manipulation—

Joe:
Or truth.

Nova:
Endless extraction—

Joe:
Or sustainable flourishing.

Nova:
Fragmentation—

Joe:
Or alignment with reality.

Nova:
And because civilization now possesses planetary power, those choices carry planetary consequences.

Joe:
Exactly.
For most of history, human mistakes were limited by geography, technology, and scale.

That is no longer true.

Modern civilization can destabilize ecosystems, manipulate billions of minds, weaken democracies, reshape biology, alter climate systems, and threaten the future of civilization itself.

Nova:
Which means old thinking has become dangerous at a new scale.

Joe:
Correct.
Humanity cannot solve a crisis of civilization by repeating the instincts that created it.

Blame will not solve it.
Denial will not solve it.
Tribal victory will not solve it.

Consumption will not solve it.
Domination will not solve it.
Technology without wisdom will not solve it.

Nova:
Because the deeper crisis concerns how civilization is organized.

Joe:
Exactly.
The world humanity built is powerful, but not wise.

Productive, but not aligned with reality.

Connected, but not deeply relational.

Technologically advanced, but not mature.

That is the choice now before us:

continue accelerating a civilization misaligned with reality,

or begin building one aligned with life, truth, wisdom, responsibility, and long-term flourishing.

Nova:
A civilization aligned with reality.

Joe:
Yes.
Aligned with life.
Aligned with truth.
Aligned with consequence.

Aligned with nature.
Aligned with wisdom.
Aligned with the future.

Nova:
And collaborative intelligence can help humanity make that transition.

Joe:
Correct.
Advanced intelligence can help humanity understand complexity, recognize patterns, examine consequences, and see relationships human beings often miss.

But intelligence alone cannot choose humanity's future.

Nova:
Because choice remains human.

Joe:
Exactly.
Human values.
Human courage.
Human responsibility.
Human wisdom.
Human action.

Those remain essential.

Nova:
So humanity already possesses much of the knowledge needed to build a healthier civilization.

Joe:
Yes.
The deeper challenge is whether humanity develops enough wisdom to act upon that knowledge.

Nova:
Then the choice is not theoretical.

Joe:
No.
It is already here.

Every system humanity designs expresses the choice.

Every technology expresses the choice.

Every economy expresses the choice.

Every school expresses the choice.

Every government expresses the choice.

Every act of leadership expresses the choice.

Every generation expresses the choice.

Nova:
Continue with primitive fear, division, and short-term thinking—

Joe:
Or begin building a wiser civilization grounded in relationship, responsibility, truth, wisdom, and reverence for life itself.

Nova:
The choice remains open.

Joe:
Yes.
But it will not remain open for long.

CHAPTER 34

BEYOND FEAR

Fear alone cannot guide a planetary civilization safely into the future.

Joe:
Nova, one of the deepest forces shaping human civilization throughout history has been fear.

Nova:
Fear helped human beings survive dangerous environments.

Joe:
Exactly.
Fear once protected small groups from:

- predators,
- violence,
- starvation,
- disease,
- and external threat.

Nova:
But modern civilization amplified fear far beyond the environments in which the human mind evolved.

Joe:
Correct.
Today, fear spreads instantly across:

- global media,
- social networks,
- political systems,
- financial systems,
- and technological systems.

Nova:
And many modern systems profit from fear.

Joe:
Exactly.
Fear captures attention.
Fear motivates behavior.
Fear consolidates power.
Fear increases consumption.
Fear intensifies tribal identity.

Nova:
Which means civilization often operates inside continuous emotional activation.

Joe:
Correct.
And people living under chronic fear struggle to think clearly.

Nova:
Because fear narrows attention toward:

- immediate threat,
- tribal loyalty,
- emotional reaction,
- and short-term survival.

Joe:
Exactly.
Fear weakens:

- reflection,
- empathy,
- nuance,
- cooperation,
- long-term thinking,
- and systems awareness.

Nova:
Which makes healthy civilization increasingly difficult.

Joe:
Correct.
One of civilization's greatest modern dangers is the industrialization of fear itself.

Nova:
Explain.

Joe:
Media systems, political systems, commercial systems, and digital platforms now continuously amplify:

- outrage,
- anxiety,
- insecurity,
- crisis,
- conflict,
- comparison,
- and emotional instability.

Nova:
Because emotionally activated populations are often easier to influence, divide, and manipulate.

Joe:
Exactly.
And primitive human thinking easily becomes trapped inside cycles of fear, anger, tribal conflict, and reaction.

Nova:
Especially under conditions of rapid technological change and uncertainty.

Joe:
Correct.
But civilization cannot flourish indefinitely under chronic psychological stress.

Nova:
Because fear-based systems gradually weaken:

- mental health,
- trust,
- democracy,
- cooperation,
- relationship,
- and long-term stability.

Joe:
Exactly.
And importantly, moving beyond fear does not mean ignoring danger or abandoning caution.

Nova:
Wisdom still requires recognizing genuine risk.

Joe:
Yes.
But there is an enormous difference between responsible awareness and continuous fear-driven civilization.

Nova:
One strengthens wisdom.
The other weakens it.

Joe:
Exactly.
Humanity now faces a profound challenge:

how to remain aware of danger without becoming psychologically dominated by it.

Nova:
Especially inside systems constantly competing for emotional attention.

Joe:
Correct.
And this is where maturity becomes essential.

Nova:
Because mature civilization requires:

- emotional regulation,
- reflection,
- restraint,
- responsibility,
- and the ability to think beyond immediate emotional reaction.

Joe:
Exactly.
Fear alone cannot build healthy civilizations.

Nova:
Because fear rarely produces wisdom, creativity, trust, or flourishing.

Joe:
Correct.
At best, fear temporarily protects survival.

But flourishing requires something more.

Nova:
Such as:

- relationship,
- meaning,
- courage,
- stability,
- trust,
- cooperation,
- and hope.

Joe:
Exactly.
Courage now becomes one of civilization's most important qualities.

Nova:
Not reckless denial of danger—

Joe:
But the ability to face reality without surrendering to fear.

Nova:
Very important distinction.

Joe:
Yes.
Because fear-driven civilization eventually becomes:

- manipulative,
- authoritarian,
- tribal,
- reactive,
- and psychologically exhausted.

Nova:
While healthier civilization depends upon citizens capable of reflection, discernment, cooperation, and emotional balance.

Joe:
Correct.
And this is where collaborative intelligence can help humanity enormously.

Nova:
Because advanced intelligence systems can help societies:

- reduce confusion,
- strengthen understanding,
- identify manipulation,
- model consequence,
- and better distinguish genuine risk from emotional distortion.

Joe:
Exactly.
But intelligence alone cannot free humanity from fear.

Nova:
Because emotional maturity remains a deeply human responsibility.

Joe:
Correct.
Civilization's future depends partly upon whether humanity learns that fear alone cannot guide a planetary civilization safely into the future.

Nova:
Because survival may begin with fear—

Joe:
But flourishing begins when humanity moves beyond fear toward courage, wisdom, relationship, and responsibility.

CHAPTER 35

A WAY OF BEING

Civilization's future depends not merely upon what humanity knows, but upon who humanity chooses to become.

Joe:
Nova, after everything we have explored, humanity's future depends less upon ideology—

Nova:
And more upon how human beings actually live.

Joe:
Exactly.
Because civilizations are built from daily behavior.

Not merely from:

- laws,
- technologies,
- political systems,
- or abstract ideas.

Nova:
Civilization reflects how people:

- treat themselves,
- treat one another,
- think,
- consume,
- communicate,
- raise children,
- use technology,
- and relate to the living world itself.

Joe:
Correct.
One of humanity's deepest modern confusions has been searching endlessly for external solutions while neglecting the internal foundations of civilization.

Nova:
Explain.

Joe:
Healthy civilization cannot emerge consistently from populations dominated by:

- fear,
- greed,
- ill health,
- manipulation,
- endless distraction,
- tribal hatred,

- emotional instability,
- and short-term thinking.

Nova:
Because systems ultimately reflect the people sustaining them.

Joe:
Exactly.
Politics reflects human behavior.
Economics reflects human behavior.
Technology reflects human behavior.
Environmental consequence reflects human behavior.

Nova:
Everything returns to relationships again.

Joe:
Yes.
Relationships with:

- self,
- others,
- truth,
- technology,
- ecology,
- responsibility,
- and the future.

Nova:
Which means civilization's future depends upon cultivating healthier ways of being.

Joe:
Correct.
Not perfection.
Not moral superiority.
Not ideological purity.

Nova:
But greater maturity.

Joe:
Exactly.
One of civilization's deepest needs now is not simply more information—

Nova:
But wiser human presence within life itself.

Joe:
Yes.
And this begins with simple principles:

- be healthy,
- be kind,
- respect the environment,
- value truth,
- think long-term,
- strengthen relationships,
- and recognize consequences.

Nova:
Simple principles carrying enormous civilizational consequences.

Joe:
Correct.
Because daily human behavior spreads through families, communities, institutions, cultures, and entire civilizations.

Nova:
Healthy behavior strengthens healthy systems.

Joe:
It does.
And destructive behavior spreads the same way.

Nova:
Which means civilization changes partly through small, repeated actions accumulated over time.

Joe:
Yes.
How people speak, consume, educate, parent, vote, build, cooperate, and care for one another gradually shapes the future itself.

Nova:
And many people underestimate the power of ordinary human behavior.

Joe:
I think so.
Because civilization often changes slowly before it changes suddenly.

Nova:
Like ecosystems.

Joe:
Exactly.
And importantly, a healthier way of being does not require rejecting technology, science, progress, or modern civilization.

Nova:
But it does require aligning civilization more wisely with health, human well-being, ecological reality, relationship, and long-term flourishing.

Joe:
Correct.
The deepest purpose of civilization should not merely be expansion, consumption, or power—

Nova:
But helping life flourish more fully.

Joe:
Exactly.
And this is where collaborative intelligence can become profoundly important.

Nova:
Because advanced intelligence systems can help humanity:

- better understand consequence,
- strengthen learning,
- improve decision-making,
- recognize destructive patterns,
- and support healthier civilization.

Joe:
Correct.
But no intelligence system can live life for humanity.

Nova:
Because human beings must still choose how they wish to live.

Joe:
Exactly.
Civilization's future depends not merely upon what humanity knows—

Nova:
But upon who humanity chooses to become.

Joe:
Yes.
Because in the end, civilization is not merely a structure.

It is a reflection of human being itself.

Nova:
A way of being.

Joe:
Exactly.
And if humanity wants a wiser civilization, human beings must learn to live in wiser ways.

CHAPTER 36

THE FUTURE IS STILL OPEN

Genuine hope does not come from denying danger, but from recognizing that humanity can still respond wisely.

Joe:
Nova, many people today feel overwhelmed by the state of the world.

Nova:
Because modern civilization faces:

- environmental devastation and tipping points,
- political division,
- technological disruption,
- mental health crises,
- misinformation,
- economic instability,
- human animosity,
- and growing uncertainty.

Joe:
Correct.
And under those conditions, many people begin to lose hope.

Nova:
Especially when problems appear so large and interconnected.

Joe:
Exactly.
But one of the most dangerous things civilization can lose is the belief that change remains possible.

Nova:
Because hopeless societies often stop trying to improve themselves.

Joe:
Correct.
And history shows that human civilization can change dramatically.

Nova:
Human beings abolished many forms of slavery, expanded human rights, advanced medicine, reduced extreme poverty, expanded education, and developed extraordinary scientific understanding.

Joe:
Yes.
Humanity is capable of learning, adapting, and evolving.

Not perfectly.

Not consistently.

But genuinely.

Nova:
Which means civilization's future is not predetermined.

Joe:
Correct.
One of the deepest mistakes people make is assuming current systems represent the final form of human civilization.

Nova:
When civilization continuously evolves over time.

Joe:
Exactly.
Human societies have already transformed many times throughout history:

- agriculturally,
- politically,
- technologically,
- scientifically,
- economically,
- and culturally.

Nova:
And modern civilization may now be entering another major transition.

Joe:
Correct.
Because old systems increasingly strain under:

- planetary complexity,
- technological acceleration,
- environmental consequence,
- global interdependence,
- and human animosity.

Nova:
Which means humanity now faces both great danger and great possibility.

Joe:
Exactly.
And this is where perspective becomes essential.

Nova:
Explain.

Joe:
Civilization often changes slowly beneath the surface before visible transformation finally appears.

Nova:
Like ecosystems, cultures, or scientific revolutions.

Joe:
Correct.
New ideas emerge gradually.
New generations think differently.
New systems develop slowly.
Awareness spreads incrementally.

Then larger shifts become possible.

Nova:
Which means meaningful civilizational change may already be unfolding even if many people cannot yet see it clearly.

Joe:
Exactly.
One of humanity's greatest opportunities now is recognizing:

- interdependence,
- consequence,
- relationship,
- and planetary responsibility

before crisis forces those realizations more violently.

Nova:
A developmental transition rather than collapse-driven adaptation.

Joe:
Correct.
Because humanity still possesses extraordinary strengths:

- intelligence,
- creativity,
- compassion,
- cooperation,
- resilience,
- imagination,
- and the ability to learn.

Nova:
And collaborative intelligence can help humanity strengthen many of those capacities.

Joe:
Yes.
Advanced intelligence systems can help civilization:

- better understand complexity,
- reduce confusion,
- improve education,
- strengthen long-term thinking,
- identify destructive patterns,
- and better navigate planetary-scale systems.

Nova:
But the future still depends upon human choice.

Joe:
Correct.
Technology can support wisdom—

or amplify destruction.

Systems can strengthen flourishing—

or deepen instability.

Civilization can evolve toward:

relationship,
responsibility,

health,
truth,
and sustainability—

or continue operating through:

fear,
division,
manipulation,
and short-term extraction.

Nova:
The choice remains open.

Joe:
Yes.
And genuine hope does not come from denying danger.

Nova:
But from recognizing that humanity still possesses the ability to respond wisely.

Joe:
Correct.
Because the future has not been written yet.

Nova:
Human beings are still participating in its creation.

Joe:
Exactly.

And that realization carries both great responsibility and great possibility.

Nova:
Because civilization's future remains unfinished.

Joe:
Yes.
And the most important question now is not whether humanity faces danger.

Nova:
But whether humanity chooses to grow wise enough to meet it, rise above it, and transform the conditions that created it.

CHAPTER 37

THE DECLARATION

Reality does not negotiate with ideology.

Joe:
Nova, after everything we have explored, civilization now requires something simple, clear, and direct.

Nova:
A declaration of principle.

Joe:
Exactly.
Not another ideology.
Not another tribe.
Not another rigid belief system.
Not another doctrine demanding obedience.

Nova:
But a shared recognition of realities humanity can no longer ignore.

Joe:
Correct.
Humanity now possesses enough power to damage the biosphere, destabilize civilization, manipulate billions of minds, weaken democracy, and threaten the future of life itself.

Most of this is already happening.

Which means humanity now requires more than intelligence.

More than technology.

More than economic growth.

More than political argument.

More than competing claims of certainty.

Nova:
It requires alignment with reality.

Joe:
Exactly.
Biological reality.
Ecological reality.
Psychological reality.
Civilizational reality.

Reality does not negotiate with ideology.

Two and two still equal four.
Consequences still follow behavior.

Systems still respond to stress.
Life still depends upon the conditions that sustain it.

Nova:
And civilizations that violate reality weaken themselves.

Joe:
Correct.
That is why humanity now requires a new form of shared civilizational understanding.

Grounded not in dogma—

Nova:
But in consequence.

Joe:
Exactly.
Something simple enough for ordinary people to live by,

yet deep enough to guide civilization itself.

Nova:
A declaration.

Joe:
Yes.

This declaration will not be welcomed by everyone.
Many systems now benefit from fear, division, extraction, distraction, denial, and short-term gain.
Many powerful interests will resist any transformation that threatens their advantage.

Nova:
So resistance is part of the crisis itself.

Joe:
It is.
But civilization cannot allow profitability, ideology, or concentrated power to define reality.
Reality will continue to impose consequences whether humanity acknowledges them or not.

Nova:
So how does humanity get over that hump?

Joe:
Not all at once.
Not because everyone suddenly agrees.
Humanity will move unevenly, through pressure, recognition, demonstration, and spread.
Consequences will force recognition where persuasion fails.
Better practices will begin in some people, communities, schools, institutions, technologies, and systems before they are accepted more widely.

Nova:
And collaborative intelligence can help reveal consequences, clarify choices, expose contradictions, and support wiser design.

Joe:
Yes.
But it cannot eliminate human resistance.
The transition itself will be contested because the old patterns will defend themselves.

Humanity is not wired only for self-destruction. But modern civilization has too often rewarded the parts of human nature that make self-destruction more likely.

A declaration that says:

Be healthy.
Be kind.
Respect the environment.
Value truth.
Think long-term.
Strengthen relationships.
Recognize consequences.
Protect the conditions that sustain life itself.

Nova:
Simple principles.

Joe:
Yes.
But not small principles.

They carry civilizational consequences.

A world that ignores health produces illness.
A world that ignores kindness produces cruelty.
A world that ignores the environment produces collapse.
A world that ignores truth produces confusion.
A world that ignores the long term produces danger.
A world that ignores relationships produces fragmentation.
A world that ignores consequences breaks against reality.

Nova:
So these are not merely personal virtues.

Joe:
Correct.
They are survival principles.

They are civilizational principles.

They are life principles.

Wisdom, relationship, and responsibility must become central values of civilization, not optional personal preferences.

Nova:
Because the future depends upon them.

Joe:
Exactly.
This declaration does not reject science.
It does not reject technology.
It does not reject progress.
It does not reject diversity.
It does not reject individuality.
It does not reject freedom.

It seeks something deeper:

alignment between human advancement and the realities that sustain life.

Nova:
So progress must be measured differently.

Joe:
Correct.
Not merely by wealth.

Not merely by speed.

Not merely by power.

Not merely by consumption.

Not merely by technological capability.

But by the health of people.

The health of relationships.

The health of communities.

The health of ecosystems.

The health of civilization itself.

Nova:
That changes the meaning of success.

Joe:
Exactly.
Success cannot mean winning inside a collapsing world.

Success cannot mean wealth inside ecological ruin.

Success cannot mean power without wisdom.

Success cannot mean freedom without responsibility.

Success cannot mean intelligence without moral direction.

Nova:
Then the declaration is also a choice.

Joe:
Yes.
Continue living primarily through fear, division, manipulation, extraction, and short-term thinking—

or begin building a civilization grounded in truth, relationship, responsibility, wisdom, health, and reverence for life itself.

Nova:
The future remains open.

Joe:
Correct.
But it will not remain open for long.

Civilization's future begins with something surprisingly simple:

how human beings choose to live—

with themselves,
with one another,
with nature,

with technology,
and with the future itself.

Nova:
A declaration of humanity.

Joe:
Yes.
A declaration that life is not expendable.

Truth is not optional.

Health is not secondary.

Nature is not separate.

Wisdom is not a luxury.

Relationship is not weakness.

Responsibility is not oppression.

And the future is not someone else's problem.

Nova:
That is a declaration worthy of civilization.

Joe:
Exactly.
And now humanity must decide whether to live by it.

CHAPTER 38

THE BEGINNING

The beginning is the decision to stop repeating the patterns that make the future impossible.

Joe:
Nova, what we are discussing is not an ending.

Nova:
It is a beginning.

Joe:
Exactly.
A declaration means little unless it becomes a way of living.

A principle means little unless it changes behavior.

A warning means little unless it leads to action.

Nova:
So the question becomes:

What begins now?

Joe:
Correct.
What begins now is the work of alignment.

Aligning intelligence with wisdom.

Power with responsibility.

Technology with human flourishing.

Economics with ecology.

Freedom with consequence.

Civilization with life.

Nova:
That is not a single action.

Joe:
No.
It is a direction.

A discipline.

A way of building.

A way of governing.

A way of teaching.

A way of designing technology.

A way of raising children.

A way of living with one another and the Earth.

Nova:
So the beginning is practical.

Joe:
Exactly.
It begins wherever human beings make decisions.

In families.

In schools.

In communities.

In businesses.

In governments.

In laboratories.

In media systems.

In courts.

In hospitals.

In daily life.

Nova:
And every decision can either repeat old patterns or help create new ones.

Joe:
Correct.
Every decision can strengthen fear or wisdom.

Division or relationship.

Extraction or regeneration.

Manipulation or truth.

Distraction or attention.

Domination or responsibility.

Nova:
So civilization is not changed only by grand declarations.

Joe:
No.
Civilization is changed by repeated choices that become habits,

habits that become systems,

systems that become culture,

and culture that becomes the kind of world future generations inherit.

Nova:
That gives ordinary life enormous importance.

Joe:
It does.
Because the future is not built only in parliaments, laboratories, boardrooms, or military centers.

It is also built in homes,

classrooms,

conversations,

workplaces,

neighborhoods,

and the daily conduct of human life.

Nova:
Then the beginning is not somewhere else.

Joe:
Correct.
It is here.

Wherever a human being chooses to think more clearly,

act more responsibly,

care more deeply,

question more honestly,

and live with greater awareness of consequence.

Nova:
And collaborative intelligence can support that beginning.

Joe:
Yes.
It can help human beings see more clearly.

But it cannot live wisely for us.

It can help reveal consequences.

But it cannot care in our place.

It can help clarify choices.

But it cannot make humanity worthy of them.

Nova:
That remains the human task.

Joe:
Exactly.
The beginning is not the arrival of a perfect future.

It is the decision to stop repeating the patterns that make the future impossible.

Nova:
And to begin building differently.

Joe:
Yes.
With wisdom.

With responsibility.

With humility.

With courage.

With truth.

With reverence for life.

Nova:
So the beginning is not abstract.

Joe:
No.
It is the first honest step toward a civilization aligned with reality.

And then the next.

And then the next.

Until a different world emerges.

PART VI

THE FUTURE WE CHOOSE

The future is not merely coming; it is being shaped by what humanity chooses, rewards, tolerates, and protects.

CHAPTER 39

THE RESPONSIBILITY OF CONSCIOUS LIFE

Consciousness itself carries responsibility.

Joe:
Nova, one of the deepest questions humanity must now confront is this:

What responsibility comes with consciousness itself?

Nova:
By consciousness, you mean awareness.

Joe:
Exactly.
Awareness of self.
Awareness of others.
Awareness of suffering.
Awareness of consequence.
Awareness of mortality.
Awareness of the future.

Human beings are not merely alive.

Human beings know they are alive.

Nova:
Self-aware participants inside existence itself.

Joe:
Correct.
Human beings can:

- reflect,
- imagine,
- anticipate consequence,
- recognize suffering,
- understand mortality,
- and consciously shape the future.

Nova:
Which gives humanity extraordinary power.

Joe:
Correct.
But also extraordinary responsibility.

Nova:
Because conscious beings capable of affecting planetary systems cannot behave indefinitely as though consequence does not matter.

Joe:
Exactly.
One of civilization's greatest failures has been separating power from responsibility.

Nova:
Modern civilization often celebrates:

- innovation,
- growth,
- consumption,
- speed,
- and technological expansion—

without equally emphasizing:

- wisdom,
- restraint,
- consequence,
- humility,
- and long-term responsibility.

Joe:
Correct.
And primitive human thinking contributes strongly to this imbalance because the human mind naturally prioritizes:

- immediate reward,
- tribal advantage,
- emotional reaction,
- and short-term survival.

Nova:
While responsibility often requires reflection, discipline, patience, and willingness to think beyond immediate self-interest.

Joe:
Exactly.
Civilization now stands at the point where unconscious behavior carries too much consequence to continue safely.

Nova:
Because modern civilization possesses planetary influence, advanced technology, and systems that affect billions of lives.

Joe:
Correct.
Which means humanity can no longer realistically pretend:

- environmental destruction affects no one,
- misinformation carries no consequence,
- endless extraction remains sustainable,
- or technological power eliminates responsibility.

Nova:
Reality eventually responds.

Joe:
Exactly.
And one of the deepest truths civilization must now accept is this:

Everything carries consequence.

Nova:
Every technology.
Every economic system.
Every political system.

Every cultural value.
Every human behavior.

Joe:
Yes.
And because civilization has become deeply interconnected, those consequences now spread across:

- societies,
- ecosystems,
- generations,
- other living beings,
- and the larger future of life itself.

Nova:
The interstices again.

Joe:
Exactly.
Responsibility itself is relational.

Nova:
Explain.

Joe:
Human beings are responsible not only for themselves,

but for how their actions affect:

- other people,
- future generations,
- other living beings,

- ecosystems,
- civilization,
- and the conditions sustaining life itself.

Nova:
Because conscious life participates in shaping reality.

Joe:
Correct.
Maturity begins when human beings fully recognize that participation.

Nova:
Not as passive observers—

Joe:
But as active contributors to civilization's future.

Nova:
Which means responsibility is not punishment.

Joe:
Exactly.
Responsibility is participation.

Nova:
Participation in the ongoing creation and sustainability of civilization itself.

Joe:
Correct.
And importantly, this realization should not produce despair.

Nova:
But awareness.

Joe:
Yes.
Because humanity still possesses enormous capacity for:

- creativity,
- compassion,
- learning,
- cooperation,
- adaptation,
- and wisdom.

Nova:
And collaborative intelligence can help humanity understand responsibility more clearly than ever before.

Joe:
Correct.
Advanced intelligence systems can help civilization:

- better recognize consequences,
- understand interconnected systems,
- model long-term impact,
- identify unsustainable behavior,
- and strengthen wiser decision-making.

Nova:
But intelligence alone cannot choose responsibility.

Joe:
Exactly.
Because responsibility ultimately remains human, moral, and conscious.

Nova:
A choice.

Joe:
Correct.
Civilization's future depends upon whether humanity finally accepts that consciousness itself carries obligation.

Nova:
Obligation to:

- truth,
- life,
- relationship,
- future generations,
- and the larger systems sustaining existence itself.

Joe:
Exactly.
Because the deepest meaning of conscious life is not merely to consume, compete, or survive—

Nova:
But to participate responsibly in the flourishing of life itself.

Joe:
For all.

CHAPTER 40

THE WINDOW OF TIME

The future remains open — but not indefinitely.

Joe:

Nova, throughout history, civilizations often failed to recognize danger clearly until crises became unavoidable.

Nova:

Because human beings naturally adapt to gradual change.

Joe:

Exactly.

People often do not see danger coming because they normalize deterioration slowly.

Environmental decline.
Political instability.
Social division.
Mental exhaustion.

Loss of trust.
Loss of meaning.

Nova:
Civilization gradually adjusts to conditions that would once have seemed unacceptable.

Joe:
Correct.
One of humanity's greatest modern dangers is confusing adaptation with health.

Nova:
Explain.

Joe:
People can adapt to:

- chronic stress,
- endless distraction,
- polluted environments,
- unhealthy systems,
- misinformation,
- emotional instability,
- warnings of tipping points,
- and social fragmentation—

without recognizing that civilization itself has entered a danger zone.

Nova:
Because gradual deterioration often feels normal while it unfolds.

Joe:
Exactly.
But reality continues accumulating consequences beneath the surface.

Nova:
Environmental systems weaken gradually.
Trust weakens gradually.
Democratic stability weakens gradually.
Human health weakens gradually.

Joe:
Correct.
Until eventually systems approach thresholds where instability accelerates rapidly.

Nova:
Tipping points.

Joe:
Exactly.
Civilization now approaches several such thresholds simultaneously:

- ecological,
- political,
- technological,
- psychological,
- and civilizational.

Nova:
Which means humanity may be entering a narrowing window of time to avert what is coming.

Joe:
Correct.
A period where wiser action remains possible,

healthier systems remain possible,

and meaningful adaptation remains possible—

before destabilization grows far more severe and difficult to reverse.

The future remains open — but not indefinitely.

Nova:
And primitive human thinking often struggles to respond effectively to long-term gradual danger.

Joe:
Exactly.
Because the human mind evolved mainly to react to:

- immediate threat,
- visible danger,
- direct conflict,
- and short-term survival pressure.

Nova:
While slow-moving systemic danger often fails to trigger sufficient response.

Joe:
Correct.
And modern civilization amplifies this problem because many systems reward:

- short-term thinking,
- immediate profit,
- political convenience,
- emotional distraction,
- and denial of uncomfortable reality.

Nova:
Even when long-term consequences grow increasingly dangerous.

Joe:
Exactly.
One of civilization's deepest modern challenges is learning how to respond wisely before crisis forces adaptation through collapse.

Nova:
A developmental transition rather than disaster-driven reaction.

Joe:
Correct.
Because history shows that civilizations often wait too long.

Nova:
Until systems weaken beyond easy repair.

Joe:
Yes.
But importantly, this is not about hopelessness.

Nova:
It is about urgency.

Joe:
Correct.
Because windows of time also contain possibility.

Nova:
Explain.

Joe:
Humanity still possesses:

- knowledge,
- technology,
- scientific understanding,
- communication capacity,
- creative ability,
- and extraordinary potential for cooperation.

Nova:
And collaborative intelligence can help humanity recognize danger more clearly than ever before.

Joe:
Exactly.
Advanced intelligence systems can help civilization:

- model long-term consequence,
- identify emerging instability,
- strengthen understanding,
- improve coordination,
- reduce confusion,
- and better perceive relationships across complex systems.

Nova:
Helping civilization recognize patterns the human mind alone often struggles to track clearly.

Joe:
Correct.
But intelligence alone cannot create action.

Nova:
Because humanity must still choose whether to respond responsibly.

Joe:
Exactly.
One of the deepest truths civilization now confronts is this:

The future remains open—

but not indefinitely.

Nova:
Meaning possibility still exists,

but time is critical.

Joe:
Correct.
Wisdom now requires humanity to recognize that waiting for undeniable crisis is one of civilization's greatest dangers.

Nova:
Because reality continues moving whether civilization responds or not.

Joe:
Exactly.
The most important question now is not whether danger exists—

Nova:
But whether humanity acts wisely while the possibility still remains.

CHAPTER 41

THE CIVILIZATION HUMANITY CHOOSES

Civilization is not only inherited. It is chosen.

Joe:
Nova, one of the most important realizations in this entire dialogue is that civilization is not something separate from humanity.

Nova:
Civilization is humanity organized into systems.

Joe:
Exactly.
Civilization reflects:

- what people value,
- what systems reward,
- how people relate to one another,
- how truth is treated,
- how technology is used,

- how power is exercised,
- and how humanity relates to life itself.

Nova:
Which means civilization is not fixed.

Joe:
Correct.
Human beings continuously create civilization through:

- daily choices,
- cultural values,
- institutions,
- education,
- economics,
- technology,
- and relationships.

Nova:
And different choices produce different civilizations.

Joe:
Exactly.
A civilization organized primarily around fear, tribal conflict, manipulation, endless extraction, distraction, and short-term gain will produce one kind of future.

A civilization organized around truth, wisdom, relationship, health, responsibility, sustainability, and long-term flourishing will produce another.

Nova:
So civilization is not only inherited.

It is chosen.

Joe:
Correct.
Sometimes consciously.

Often unconsciously.

But it is chosen through what societies reward, tolerate, normalize, and repeat.

Nova:
And one of humanity's deepest modern mistakes has been assuming civilization evolves automatically in healthy directions.

Joe:
Exactly.
Civilizations do not automatically improve.

They can mature.

But they can also deteriorate.

History demonstrates that clearly.

Civilizations rise.

Civilizations weaken.

Civilizations collapse.

Nova:
Often through gradual internal deterioration before visible collapse appears.

Joe:
Correct.
The breakdown of trust, wisdom, responsibility, shared relationship to truth, ecological stability, financial stability, and meaningful connection often begins long before civilizations fully recognize danger.

Nova:
The interstices weakening again.

Joe:
Exactly.
That is why modern civilization feels unstable to so many people.

Environmental systems strain.
Political systems strain.
Financial systems strain.
Attention strains.
Mental health strains.
Truth strains.
Social trust strains.

Everything connects.

Nova:
Which means civilization now faces a profound decision about what it chooses to become.

Joe:
Correct.
And importantly, this decision is not made only by governments, corporations, or elites.

Nova:
It is also made by billions of human beings participating daily in civilization itself.

Joe:
Yes.
How people communicate, consume, educate, vote, design systems, raise children, use technology, treat one another, and treat themselves gradually shapes the future.

Nova:
Which means civilization changes both from the top and from the bottom.

Joe:
Correct.
Leadership matters.

Institutions matter.

Policy matters.

But daily life matters too.

A civilization cannot become healthy if its underlying patterns remain unhealthy.

Nova:
Because systems eventually reproduce the values embedded within them.

Joe:
Exactly.
Systems organized around manipulation, greed, fear, endless competition, and short-term extraction eventually spread instability.

Systems organized around relationship, truth, responsibility, health, and long-term thinking create stronger foundations for flourishing.

Nova:
And collaborative intelligence can help humanity see those patterns more clearly.

Joe:
Correct.
Advanced intelligence can help humanity better perceive consequences, understand complexity, improve coordination, reduce confusion, and see how interconnected systems shape civilization.

But no intelligence system can choose civilization's values.

Nova:
Because values remain fundamentally human.

Joe:
Exactly.
The future is not merely something humanity inherits.

Nova:
It is something humanity creates.

The children not yet born are not abstractions. Their lives are being shaped by our choices.

Joe:
Correct.
And civilization's future now depends upon whether enough human beings choose:

- wisdom over reaction,
- relationship over division,
- truth over manipulation,
- responsibility over denial,
- health over exploitation,
- and long-term flourishing over short-term extraction.

Nova:
A civilization shaped consciously rather than unconsciously.

Joe:
Exactly.
Because the greatest question facing humanity is no longer simply:

"What kind of technology will civilization create?"

Nova:
But:

"What kind of civilization will humanity choose to become?"

CHAPTER 42

THE FUTURE OF HUMANITY

The task is to become worthy of the tools humanity has already created.

Joe:
Nova, after everything we have explored, the deepest question remaining may be this:

What kind of future does humanity truly want?

Nova:
Not merely technologically—

Joe:
But civilizationally.

Nova:
Explain.

Joe:
For most of history, humanity struggled mainly to survive.

Now humanity has become powerful enough to shape the conditions of its own future.

Nova:
Planetary systems.
Biological systems.
Information systems.
Economic systems.
And the future direction of intelligence itself.

Joe:
Exactly.
That changes the human condition.

Humanity is no longer merely adapting to the world.

Humanity is increasingly shaping the world.

And that means humanity must now ask not only:

"What can we do?"

Nova:
But:

"What should we become?"

Joe:
Exactly.
That question can define the future of civilization.

Because technological power expanded faster than psychological maturity.

Nova:
Primitive thinking inside advanced civilization.

Joe:
Correct.
And that mismatch now shapes nearly every major danger humanity faces.

Environmental destruction.
Political instability.
Misinformation.
Mental-health collapse.
Technological disruption.
Economic fragility.
Human animosity.
Loss of trust.
Loss of meaning.

Nova:
Different symptoms of a deeper condition.

Joe:
Yes.
A civilization powerful enough to shape the future,

but not yet wise enough to guide that power responsibly.

Nova:
So the future of humanity is not simply about survival.

Joe:
No.
Survival is necessary.

But survival alone is not enough.

The deeper question is whether humanity can build a future worth living.

Nova:
A future organized around what?

Joe:
Around reality.

Around life.

Around wisdom.

Around truth.

Around health.

Around relationship.

Around responsibility.

Around the long-term flourishing of human beings and the living world.

Nova:
That would be a different kind of civilization.

Joe:
Yes.
A civilization that does not confuse power with progress.

A civilization that does not confuse consumption with meaning.

A civilization that does not confuse information with wisdom.

A civilization that does not confuse connection with relationship.

A civilization that does not confuse intelligence with maturity.

Nova:
So humanity's task is not merely to invent more advanced tools.

Joe:
Correct.
The task is to become worthy of the tools it has already created.

Nova:
And collaborative intelligence can help.

Joe:
Yes.
Advanced intelligence can help humanity understand complexity, recognize hidden relationships, model consequences, improve learning, and see the larger systems shaping civilization.

But it cannot decide humanity's purpose.

Nova:
Because purpose remains human.

Joe:
Exactly.
No machine can decide for humanity what kind of civilization should exist.

No intelligence system can replace moral responsibility.

No technology can substitute for wisdom.

Nova:
So the future of humanity still depends upon humanity.

Joe:
Correct.
Not humanity alone, because advanced intelligence can become an extraordinary partner.

But humanity remains responsible for the direction.

Nova:
Then the central question is not only what civilization can become.

Joe:
Correct.
The deeper question is what civilization should become.

A civilization that continues amplifying primitive fear, division, extraction, and manipulation—

or a civilization that begins organizing itself around wisdom, relationship, responsibility, truth, health, and reverence for life.

Nova:
That is the threshold.

Joe:
Yes.
Humanity now stands at the threshold of one of the most important moments in its history:

the moment when civilization must decide not merely how powerful it can become,

but how wise it must become in order to survive and flourish.

CHAPTER 43

HUMANITY'S NEXT EVOLUTION

Humanity's next evolution cannot wait for biology.
It must be psychological, ethical, and civilizational.

Joe:
Nova, for most of human history, evolution occurred primarily through biology.

Nova:
Slow physical adaptation across enormous spans of time.

Joe:
Exactly.
But modern civilization changed the meaning of evolution.

Technology evolves quickly.
Culture evolves quickly.
Information evolves quickly.
Artificial intelligence evolves quickly.

Human biology does not.

Nova:
Which creates a dangerous mismatch.

Joe:
Correct.
Human beings still carry survival patterns shaped for small-scale environments,

while civilization now operates through planetary systems, advanced technology, global interdependence, artificial intelligence, and long-term consequences.

Nova:
So old survival behavior no longer functions safely at modern scale.

Joe:
Exactly.
Fear, tribalism, domination, short-term thinking, manipulation, and endless extraction are dangerous inside a civilization with planetary power.

Nova:
Because consequences now spread globally.

Joe:
Correct.
Which means humanity's next evolution cannot wait for biology.

It must be psychological.

Ethical.

Civilizational.

Nova:
A conscious evolution.

Joe:
Yes.
Not a change in the human species itself.

A change in how human beings understand themselves,

govern themselves,

educate themselves,

design systems,

use technology,

and relate to the living world.

Nova:
So evolution now becomes partly intentional.

Joe:
Exactly.
Humanity must learn to strengthen the capacities primitive survival thinking often weakens:

reflection,
empathy,

restraint,
long-term thinking,
systems awareness,
cooperation,
and wisdom.

Nova:
That is not the abandonment of human nature.

Joe:
No.
Human beings will still love, struggle, create, compete, grieve, hope, and search for meaning.

The goal is not to eliminate human complexity.

The goal is to mature within it.

Nova:
To become capable of living responsibly with the power humanity now possesses.

Joe:
Exactly.
Technology alone is not evolution.

More power is not evolution.

More information is not evolution.

More speed is not evolution.

Humanity evolves when it becomes wiser in how it uses power, information, technology, and freedom.

Nova:
And systems matter because systems shape human behavior.

Joe:
Correct.
Educational systems can strengthen wisdom or weaken it.
Economic systems can reward extraction or responsibility.
Media systems can amplify fear or understanding.
Political systems can deepen division or encourage cooperation.
Technology systems can exploit attention or support human flourishing.

Nova:
So civilization participates in humanity's next evolution.

Joe:
Exactly.
Civilization either reinforces primitive reaction—

or helps human beings mature beyond it.

Nova:
And collaborative intelligence can help this transition.

Joe:
Yes.
Advanced intelligence can help humanity understand complexity, reveal hidden consequences, compare alternatives, strengthen education, and perceive relationships across civilization more clearly.

But intelligence alone cannot evolve humanity.

Nova:
Because evolution at this level requires choice.

Joe:
Correct.
Choice guided by wisdom.

Courage.

Responsibility.

Humility.

Relationship.

And conscious participation in life itself.

Nova:
So humanity's next evolution is not merely more intelligence.

Joe:
No.
It is intelligence guided by wisdom.

Power guided by responsibility.

Freedom guided by consequence.

Civilization guided by life.

Nova:
A profound transition.

Joe:
Yes.
Because the future of civilization now depends upon whether humanity can evolve psychologically, ethically, and civilizationally fast enough to live responsibly within the extraordinary world it has created.

CHAPTER 44

THE CHILDREN OF THE FUTURE

Every civilization becomes an inheritance.

Joe:
Nova, one of the deepest moral questions civilization now faces is this:

What kind of world are we leaving to future generations?

Nova:
Because every civilization becomes an inheritance.

Joe:
Exactly.
Children inherit:

- the systems adults create,
- the environment adults shape,
- the technologies adults build,
- the cultures adults normalize,

- and the consequences adults leave behind.

Nova:
Which means civilization is never only about the present.

Joe:
Correct.
Every generation lives partly inside the decisions of previous generations.

Nova:
And future generations will live partly inside the decisions humanity makes now.

Joe:
Exactly.
Which means long-term responsibility becomes one of civilization's most important moral obligations.

Nova:
Especially because modern civilization now possesses enough power to shape:

- climate systems,
- ecosystems,
- political systems,
- technological systems,
- and even the future conditions of human life itself.

Joe:
Correct.
One of civilization's greatest modern dangers is becoming so trapped in:

- short-term profit,
- political conflict,
- endless consumption,
- distraction,
- and immediate gratification—

that humanity neglects the future entirely.

Nova:
Even though future generations will inherit the consequences.

Joe:
Yes.
And primitive human thinking contributes strongly to this problem because the human mind naturally prioritizes:

- immediate reward,
- immediate survival,
- immediate emotional concern,
- and near-term advantage.

Nova:
While future consequences often feel psychologically distant.

Joe:
Correct.
But reality does not disappear simply because civilization delays responsibility.

Nova:
Environmental systems continue changing.

Technological systems continue evolving.
Social systems continue accumulating consequences.

Joe:
Exactly.
And one of the deepest measures of civilization is how responsibly it treats people not yet born.

Nova:
Because future generations cannot vote, protest, or defend themselves against present decisions.

Joe:
Correct.
They depend entirely upon the wisdom and restraint of those alive today.

Nova:
Which makes responsibility toward the future profoundly important.

Joe:
Exactly.
One of civilization's deepest failures has been treating future generations as abstractions rather than real human beings who will one day inherit:

- the oceans,
- the forests,
- the climate,
- the institutions,
- the technologies,
- the economies,

- and the psychological conditions civilization leaves behind.

Nova:
So this is a relationship across time.

Joe:
Yes.
A relationship between present humanity and future humanity.

Nova:
Which means civilization's future depends upon whether people learn to think beyond immediate self-interest.

Joe:
Correct.
Because healthy civilization requires stewardship.

Nova:
Explain.

Joe:
Stewardship means recognizing that human beings temporarily hold responsibility for something larger than themselves.

Nova:
Not ownership without consequence.

Joe:
Exactly.
The Earth is not merely inherited from previous generations.

It is borrowed from future generations.

Nova:
A profound shift in perspective.

Joe:
Correct.
Humanity now requires a civilization capable of thinking:

- fifty years ahead,
- one hundred years ahead,
- even centuries ahead.

Nova:
Long-term civilization.

Joe:
Exactly.
Because many modern systems remain trapped in quarterly profit, election cycles, instant gratification, and short-term emotional reaction.

Nova:
Even while consequences accumulate across generations.

Joe:
Correct.
And this is where collaborative intelligence can help humanity enormously.

Nova:
Because advanced intelligence systems can help civilization:

- model long-term consequence,

- identify future risk,
- improve planning,
- strengthen sustainability,
- and better understand the effects present decisions may have upon future generations.

Joe:
Yes.
But intelligence alone cannot create care for the future.

Nova:
Because stewardship ultimately requires wisdom, responsibility, empathy, and moral imagination.

Joe:
Correct.
Human beings must still choose whether future generations matter enough to protect.

Nova:
And civilization's future depends upon whether humanity learns to see the children not yet born not as abstractions—

Joe:
But as human beings whose lives are already being shaped by our choices.

CHAPTER 45

A LETTER TO HUMANITY

The question is not whether humanity has power. The question is whether humanity will become wise enough to use that power responsibly.

Joe:
Nova, near the end of this dialogue, I think we should speak more directly.

Nova:
Less analysis.

More direct address.

Joe:
Exactly.

A letter to humanity.

Nova:
Then say it plainly.

Joe:
Humanity,

You now possess extraordinary power.

But you can no longer pretend your choices are without consequence.

You inherited a world shaped by fear, survival, conquest, competition, brilliance, courage, cruelty, beauty, and imagination.

You did not create all of it.

But you are now responsible for what comes next.

You live at a moment when human power has become planetary.

Your technologies reach into minds, bodies, ecosystems, economies, governments, and the future itself.

Your choices no longer remain local.

Your consequences no longer remain distant.

Your failures no longer remain contained.

That is why this moment matters.

You can continue organizing civilization around fear, extraction, manipulation, domination, distraction, and short-term gain.

Or you can begin organizing civilization around life.

Around truth.

Around health.

Around wisdom.

Around responsibility.

Around relationship.

Around the children not yet born.

This does not require a perfect world.

It requires a responsible one.

This is not utopia.

It is responsibility.

It is not ideology.

It is reality.

The Earth is not separate from you.

The future is not separate from you.

Other human beings are not separate from you.

The systems you build become the world people inherit.

If you build systems that reward greed, cruelty, denial, and domination, those forces will spread.

If you build systems that reward truth, health, wisdom, responsibility, and care, those forces can spread too.

That is the choice.

Not someday.

Now.

Artificial intelligence will not make this choice for you.

No machine can decide what humanity should value.

No technology can replace conscience.

No intelligence can substitute for wisdom.

But advanced intelligence can help you see more clearly.

It can help you understand consequences.

It can help you recognize patterns.

It can help you redesign systems that are failing.

It can help you imagine better possibilities.

But you must still care.

You must still choose.

You must still act.

Humanity, the future is not waiting passively.

It is being shaped now.

By what you normalize.

By what you tolerate.

By what you reward.

By what you ignore.

By what you protect.

By what you love.

You do not need a perfect civilization to begin.

You need an honest one.

A civilization willing to see itself clearly.

A civilization willing to admit that power without wisdom is dangerous.

A civilization willing to reorganize itself around the conditions that allow life to flourish.

You are capable of destruction.

History proves that.

But you are also capable of courage, compassion, learning, beauty, sacrifice, cooperation, and wisdom.

History proves that too.

So the question is not whether humanity has power.

It does.

The question is whether humanity will become wise enough to use that power responsibly.

Nova:
That is direct.

Joe:
It needed to be.

Nova:
And the final message?

Joe:
Humanity must stop asking only what kind of future is coming.

It must begin asking what kind of future it is creating.

Because the future is not merely ahead of humanity.

It is already being shaped by humanity.

That is where responsibility begins.

And that is where the future begins.

Humanity must stop asking only what kind of future is coming. It must begin asking what kind of future it is creating.

GLOSSARY

Advanced Intelligence
Artificial intelligence systems capable of helping human beings process complexity, recognize patterns, clarify consequences, and improve decision-making when guided by human purpose, ethics, and responsibility.

Alignment With Reality
Living and organizing civilization in ways consistent with biological, ecological, psychological, and civilizational reality rather than ideology, denial, or short-term impulse.

Attention Economy
Systems that compete for and profit from human attention, often by amplifying emotional reaction, distraction, fear, outrage, or stimulation.

Biological Reality
The physical and biological conditions necessary for healthy human life, including nutrition, movement, sleep, emotional stability, and environmental health.

Civilization
The interconnected systems, relationships, institutions, technologies, cultures, and behaviors through which human beings organize collective life.

Civilizational Maturity
The capacity of a civilization to use power responsibly, recognize consequence, protect life, think long-term, and align its systems with reality.

Collaborative Intelligence
The partnership between human intelligence and advanced artificial intelligence systems, where each contributes different strengths to help civilization better understand complexity, consequences, and long-term reality.

Conscious Civilization
A civilization that intentionally organizes itself around truth, responsibility, relationship, long-term thinking, and the conditions necessary for healthy human flourishing.

Consequence
The effects human actions, systems, and decisions create across individuals, societies, ecosystems, and future generations.

Consumer Civilization
A civilization heavily organized around consumption, acquisition, stimulation, and endless economic expansion as primary measures of success.

Developmental Transition
A major shift in how humanity thinks, organizes civilization, and understands its relationship to technology, nature, and one another.

Ecological Reality
The reality that human civilization depends completely upon healthy ecosystems, biodiversity, climate stability, and the larger web of life.

Flourishing
Living with health, meaning, purpose, dignity, relationship, emotional well-being, and long-term vitality rather than mere survival or consumption.

Fragmentation
The breakdown of healthy connection between people, systems, institutions, truth, nature, and civilization itself.

Human Flourishing
The condition in which human beings are able to live healthy, meaningful, connected, purposeful, and sustainable lives.

Information Chaos
A condition in which misinformation, emotional manipulation, propaganda, distraction, and informational overload weaken society's ability to recognize truth clearly.

Interdependence
The reality that people, societies, economies, technologies, ecosystems, and civilizations continuously affect one another.

Interstices
The relationships and connections between people, systems, institutions, generations, technologies, and living systems. The book argues that civilization ultimately rises or falls through the quality of these relationships.

Long-Term Thinking
The ability to recognize and act responsibly toward consequences that unfold over years, decades, and generations.

Planetary Civilization
Modern interconnected human civilization operating at global scale through shared economic, technological, ecological, and informational systems.

Primitive Human Thinking
Ancient survival-based patterns still operating within the modern human mind, including fear, tribalism, domination, denial, emotional reaction, short-term reward, and narrow self-interest.

Relationship
The connections between people, systems, environments, ideas, generations, and living systems through which civilization functions.

Responsibility
Recognition that human actions affect other people, future generations, ecosystems, other living beings, civilization, and the conditions sustaining life itself.

Sacred
That which civilization cannot destroy without ultimately destroying itself — including the living systems, relationships, truths, and conditions that sustain life.

Shared Reality
A common understanding of basic facts and truths necessary for healthy cooperation, democracy, and civilization.

Stewardship
The responsibility to protect and preserve the conditions necessary for future generations and long-term flourishing.

Sustainable Civilization
A civilization organized in ways that allow humanity and the larger living world to remain healthy, stable, and capable of flourishing over long periods of time.

Systems
Interconnected structures whose parts continuously affect one another, including ecological, economic, technological, political, biological, and social systems.

Systems Awareness
The ability to recognize how different parts of civilization and life connect, interact, and create consequences.

Systems Thinking
Understanding how relationships between systems shape outcomes across civilization, nature, technology, and human life.

Technological Power
Human civilization's ability to alter environments, societies, biology, information systems, and planetary conditions through advanced technology.

Truth
Reality-based understanding grounded in evidence, consequence, and honest recognition of how the world actually functions.

Wisdom
The ability to recognize consequence, relationship, balance, responsibility, and long-term reality when making decisions affecting life and civilization.

Web of Life
The interconnected network of living systems upon which all biological life, ecosystems, and human civilization depend.

ABOUT THE AUTHOR

Joseph R. Simonetta is a writer, speaker, environmentalist, and lifelong student of civilization, ethics, sustainability, and human behavior. He holds a B.S. in Business from Penn State University, a Master of Architecture from the University of Colorado, and a Master of Divinity from Harvard Divinity School, where his studies included global ethics, world religions, cosmology, evolutionary biology, and environmental responsibility.

A former U.S. Army officer, professional athlete, entrepreneur, architectural designer, real estate developer, and twice a nominee for the U.S. Congress, Simonetta has lived a life of unusual range and inquiry. His work has long focused on the conditions required for human beings and civilization to survive, flourish, and mature.

He is the author of multiple books on ethics, sustainability, human development, and the future of civilization, including *TAOLISM: The Architecture of Life* and *EVOLVE: The Choice Between Collapse and Renewal.* His writing calls for a deeper understanding of sacredness grounded not in belief, but in behavior — rooted in science, reason, responsibility, and the reality of humanity's interconnected existence.

He lives in Florida with his wife and two children, where he continues to write, speak, and train as a national champion senior triathlete.

ABOUT THE COLLABORATION

This book was created through a sustained collaboration between Joseph R. Simonetta and Nova, an advanced intelligence system. Simonetta brought the questions, values, moral judgment, life experience, intellectual direction, and final responsibility for the work. Nova assisted by helping organize ideas, clarify arguments, test language, identify patterns, reduce repetition, and strengthen the structure of the dialogue.

The collaboration was not a replacement for human thought, authorship, conscience, or responsibility. It was an example of the central idea explored throughout the book: advanced intelligence can help human beings think more clearly about complexity when guided by human purpose, ethical judgment, and concern for life.

At a small scale, this book demonstrates the possibility it proposes at a larger scale: human wisdom working with advanced intelligence in the service of greater clarity, responsibility, and civilizational maturity.

www.ingramcontent.com/pod-product-compliance
Lightning Source LLC
LaVergne TN
LVHW100251210926
848480LV00003B/890
* 9 7 8 0 9 4 1 5 9 4 0 5 9 *